WHAT IN THE W🦠RLD ARE VIRUSES

What In The World Are
VIRUSES

WRITTEN BY
CASSANDRA VANDRUNEN-LACHANCE, YI YANG FEI, TIM CHAPMAN, RICO CUECACO, CHITRINI TANDON, CAMRYN KABIR-BAHK, CHRISTINA NGUYEN, BRIANNA BEDRAN, NOAH VARGHESE

EDITED BY GRACE NZOVWA ZULU-SITIMA

First Printing: 2021

Typeset and Cover Design by Michelle Wong

ISBN 978-1-77369-252-4

Golden Meteorite Press
103 11919 82 St NW
Edmonton, AB T5B 2W3
www.goldenmeteoritepress.com

GM
PRESS

TABLE of CONTENTS

WHAT IS THE HISTORY OF VIRUSES?

CASSANDRA VANDRUNEN-LACHANCE

In human history, different viruses have evolved and affected global populations in many unique ways. A virus can be defined as a small, infectious agent of simple composition that can only multiply within the living cells of animals, plants or bacteria (Wagner & Krug, 2020). As of 2012, 219 virus species have been identified that can infect human populations (Woolhouse et al., 2012). On top of this, new types of human viruses are being discovered at a rate of 3-4 viruses each year, composing two-thirds of all new human pathogens discovered (Woolhouse et al., 2012). Viral particles can be found nearly everywhere, including in the wilderness, urban areas, varying surfaces, within biological fluids, and within the air itself (Labadie et al., 2020). Many of these viruses that have impacted the globe have altered or changed the world we live in (Oldstone, 2020, p.3). Specific viruses, such as the yellow fever virus, are key milestone viruses in history in which new discoveries, epidemics, or events occurred. These viruses outline the most integral moments in the history of viruses.

Yellow Fever

The ailment known as yellow fever, which is caused by yellow fever virus, is known as one of the most dangerous infectious diseases (Douam & Ploss, 2018). Some of the most common symptoms of yellow fever include fever, muscle pain, headache, and nausea (World Health Organization, 2019). This virus originated in Africa but was brought into the Western hemisphere during the slave trade era, where the first epidemic reported occurred in 1648 in the region known as Yucatan (Staples & Monath, 2008). During the following 200 years, many outbreaks occurred in tropical America, the North American coastal cities, and Europe (Staples & Monath, 2008). It was in the 19th century when it was discovered that the yellow fever virus was not communicable from person-to-person, however, it was wrongly believed that the disease was caused by atmospheric miasmata (Staples & Monath, 2008). In 1881, a man named Carlos Finlay from Cuba suggested that it was mosquitoes (specifically Aedes aegypti) that spread the yellow fever virus (Staples & Monath, 2008). While Finlay was unable to prove his theory, it served as the basis for Walter Reed's research, who was able to prove that the virus was spread through these mosquitoes, which influenced American General William Gorgas to implement a campaign within Havana against this specific vector (Staples and Monath, 2008). While the elimination of Aedes aegypti in certain regions was able to provide some positive effects in the eradication of yellow fever, the long-term and widespread elimination of this ailment was dispelled by the discovery that yellow fever is a zoonosis that is maintained by sylvatic mosquito species and non-human primates from the Amazon jungle (Staples and Monath, 2008).

By the year 1930, virologist Max Theiler was able to successfully attenuate the virus in a convenient mouse animal model (Staples and Monath, 2008). Seven years later, Theiler developed the 17D strain of the yellow fever virus, which allowed for the creation of a vaccine to protect against the virus (Oldstone, 2020, p.120). Theiler was later awarded the Nobel prize in 1951 for "discoveries concerning yellow fever and how to combat it" (Oldstone, 2020, p.120). Large vaccination campaigns during the 1940-1950's and 2000's were able to help circumvent yellow fever outbreaks (Douam & Ploss, 2018). However, due to decreases in vaccination coverage in endemic areas between the 1960 and 2000's, there was an upsurge of outbreaks in the regions of South America and Africa (Douam & Ploss, 2018). In 2016, the World Health Organization(WHO) developed the Eliminate Yellow Fever Epidemics (EYE) Strategy with intent to protect all at-risk populations, prevent the international spread of the yellow fever virus and contain outbreaks quickly (World Health Organization, 2019). The World Health Organization believes that by utilizing the EYE strategy, more than 1 billion people will be protected from the yellow fever virus through vaccination by 2026 (World Health Organization, 2019).

Smallpox

Smallpox, which is known for killing nearly 300 million people in the twentieth century alone, is an important infectious disease to study (Oldstone, 2020, p.37). Caused by the Variola virus, which is part of the orthopoxvirus genus, smallpox is characterized by smallpox lesions which are bumps found upon a patient's skin (Moore et al., 2006). Additional symptoms can include headache, backache, chills, abdominal pain, and vomiting (Moore et al., 2006). The disease dates back to Pharaoh Ramses V, who is believed to have died from smallpox in 1157 BCE (Breman, 2021). It is also believed by many historians that smallpox was the cause of the plague in Athens in 430 BCE as well as the Antonine Plague that occurred from 165 to 180 AD, spreading to Europe no later than the 6th century (Greenspan, 2020). It was not until the late 1700's when the creator of the first smallpox vaccine Edward Jenner made true strides in the eradication of smallpox (Dworetzky et al., 2003). Jenner found that by inoculating an individual with cowpox, a disease that is closely linked to smallpox, one could garner the ability to protect oneself from smallpox (Dworetzky et al., 2003). This was demonstrated in 1976 by inoculating 8 year old James Phipps with material from a sore on the hand of Sarah Nelmes, a milkmaid who had contracted cowpox (Dworetzky et al., 2003). When exposed to smallpox later on, Jenner found that "no disease followed" (Dworetzky et al., 2003). Many years later in 1967 the World Health Organization began a campaign to be rid of smallpox once and for all after advancements in vaccines (Greenspan, 2020). The World Health Organization declared smallpox eradicated in 1980, the last case occurring in Somalia in 1977 (Moore et al., 2006).

Measles

The measles virus is known as one of the most contagious viruses in human history (Griffin & Oldstone, 2009, p.1). Before the introduction of a measles vaccine, measles was the cause of millions of deaths annually across the globe (Moss & Griffin, 2012). One of the characteristic signs of measles is the onset of a rash, usually beginning on the face before spreading (Moss & Griffin, 2012). Measles has been found to date back as far as the 9th century, noted by Peresian Doctor Rhazes (Haelle, 2019). It wasn't until 1757 that the Scottish doctor Francis Home discovered that measles was caused by pathogens and made the first attempt at creating a vaccine (Haelle, 2019). In 1954, a research team under the guidance of John Enders were able to isolate the virus in a cell culture (Griffin & Oldstone, 2009, p.3). By 1958, the first measles vaccine was created; however, fellow scientist Maurice Hilleman described it as "toxic as hell" due to it causing fevers and rashes in certain patients, some having "fevers so high that they had seizures." (Haelle, 2019). Several years later in 1963, the live-attenuated vaccine was perfected to protect against measles (Griffin & Oldstone, 2009, p.3). With the development of measles vaccines, the mortality rate of measles has decreased, with 164, 000 deaths occurring in 2008 (Moss & Griffin, 2012). While measles outbreaks should be no problem with countries with strict vaccination schedules, Third World countries who lack vaccines remain susceptible, causing measles to be one of the top 10 most important causes of death from infectious diseases (Oldstone, 2020, p.126).

Influenza Virus

The Influenza virus does not refer to one single virus but a group of viruses of which the strains are constantly changing (Francis et al., 2019). The virus family can be divided into 4 types, with Influenza types A and B circulating in humans to cause seasonal epidemics and occasional pandemics (Francis et al., 2019). While type A can be divided into various subtypes based on the surface proteins it expresses, hemagglutinin (18 subtypes) and neuraminidase (11 subtypes), type B viruses belong to one of two subtypes, B/Yamagata and B/Victoria (Francis et al., 2019). The influenza viruses can be mutated through two different mechanisms: antigenic drift and antigenic shift (Francis et al., 2019). Antigenic drift refers to small changes in the viral genetic material that lead to antigenic changes over time while antigenic shift refers to abrupt changes that come from genomic re-arrangements between two or more influenza viruses (Francis et al., 2019). The symptoms of influenza are similar to those of the common cold but more severe, including fever, cough, runny or stuffy nose and severe malaise (History, 2020). According to the World Health Organization, 3-5 million cases of severe infection as well as 290, 000 - 650, 000 deaths occur each year (History, 2020). One of the earliest reports of influenza dates back 410 BCE in Northern Greece by Hippocrates (History, 2020). In the past 135 years, there have been influenza pandemics in the years 1889, 1918, 1957, 1968, 1977, and 2009 (Taubenberger & Kash, 2010).

The 'Spanish' influenza pandemic that occurred between 1918-1919 resulted in the deaths of an estimated 50 million or more individuals (Taubenberger & Kash, 2010). Later through reconstruction of the viral genome it was determined that this pandemic was caused by an H1N1 virus variant (Taubenberger & Kash, 2010). More American soldiers died due to the virus than from the battles of World War I (History, 2020). In 1957, the "Asian' pandemic was caused by a H2N2 virus variant which resulted in the deaths of 1.1 million people and the 'Hong Kong' flu of 1968 that was caused by H3N2 killed 1 million people (History, 2020). In 2009, a new influenza H1N1 virus, termed the 'Swine Flu', originating in North America spread, leading to the deaths of 203, 000 people (History, 2020). While vaccines can be used to protect individuals from influenza, scientists are tasked with hitting a 'moving target' as they need to determine how the following year's virus will mutate (History, 2020).

Zika Virus

The mosquito-borne Zika virus, which belongs to the Flavivirus genus, originates from the Zika forest in Uganda, being first identified in sentinel rhesus monkeys in 1947 (Wikan & Smith, 2016). However, the first characterization of human disease following Zika virus infection happened in 1954 in the country of Nigeria (Weaver et al., 2016). A Zika virus infection includes signs and symptoms of low-grade fever, maculopapular rash, arthralgia, and conjunctivitis (Weaver et al., 2016). While most cases of Zika virus are mild or obscure, 15-20% of those afflicted with the virus present the serious form of the disease (Oldstone, 2020, p. 282). For the first 60 years following its discovery, Zika virus was confined to the equatorial zone across Africa and Asia (Song et al., 2017). However, in 2007 the Zika virus spread to Yap Island, later spreading eastward to French Polynesia and various other pacific islands in 2013-2014 before reaching Latin America in 2015 and North America in 2016 (Song et al., 2017). Additionally, in Brazil in 2015, the first association between Zika virus and microcephaly (baby born with a smaller than normal sized head) (World Health Organization, 2018). Zika virus infection is also associated with other congenital malformations (World Health Organization, 2018). Currently, no treatment for the Zika virus exists; however, the World Health Organization outlines their plan for research, prevention, and monitoring of the Zika virus in the Zika Strategic Response Framework (World Health Organization, 2018).

Human papillomavirus

Human papillomavirus is the number one most common sexually transmitted infection in the United States of America as well as being a major cause of cervical cancer (Mehta et al., 2012). In fact, there are more than 200 different genotypes of human papillomavirus that have been identified (de Sanjosé, 2018). While most of these forms cause no symptoms and others cause verrucas and warts, the link between certain strains of human papillomavirus and cancer was first noticed in the 1950's and 60's (Smith, 2014). Researchers compared the lifestyles of different women and found that cervical cancer appeared to be more common in the women who had started having intercourse at a younger age or had more sexual partners (Smith, 2014). Scientists found this odd as cancer itself is not contagious (Smith, 2014). This observation interested Harold zur Hausen, who through extensive research discovered in 1983 that Human papillomavirus 11 was present in 3 out of 24 cervical cancer samples he tested (Smith, 2014). He additionally discovered later that human papillomavirus 16 was present in half of cervical cancers while human papillomavirus 18 is present in 1 out of 5 cases (Smith, 2014). It is now known that human papillomaviruses 16 and 18 account for approximately 70% of all cervical cancer patients (Crosbie et al., 2013). By discovering this link, zur Hausen was awarded the Nobel prize (Smith, 2014). In 2006, a quadrivalent vaccine containing major capsid proteins of human papillomavirus strains 6, 11, 16, and 18 known as Gardasil®

became the first vaccine to be used to prevent cervical cancer and genital warts in women aged 9-26 years old (Shi et al., 2007). Later, in 2009, a second bivalent vaccine known as Cervarix entered the American market (Mehta et al., 2012).

HIV (human immunodeficiency virus)/AIDS (acquired immunodeficiency syndrome)

Throughout the 1980's and 1990's an HIV/AIDS epidemic occurred throughout America and the rest of the globe (Volberding, 2017). HIV is a type of virus that attacks the immune system, specifically the CD4 cells (History, 2021). If too many CD4 cells are destroyed, the body becomes unable to fight off infections or diseases, leading to AIDS which is the most severe form of an HIV infection (History, 2021). HIV can be transmitted through many bodily fluids such as blood, breast milk, semen and vaginal secretions as well as from mother to fetus during pregnancy and delivery (World Health Organization, 2020). The origin of HIV has been traced back to the simian immunodeficiency virus (SIV) in monkeys (Oldstone, 2020 p. 190). While HIV was believed to have traveled to America in the 1970's, public attention was not garnered until 1981 when Centers for Disease Control and Prevention (CDC) released a report describing how 5 previously healthy homosexual men became infected with Pneumocystis pneumonia, a normally harmless fungus that rarely affects individuals who have normal immune system function (History, 2021). The next year, the New York times published an article about a new immune system disorder that gay-related immune deficienc (GRID) due to the fact it appeared mostly in homosexual men (History, 2021). At this time, 335 people had been infected, 136 of them dying from it (History, 2021). By 1985, there were more than 20, 000 cases of AIDS reported, with a minimum of one case per global region (History, 2021). It wasn't until 1987 that zidovudine, also known as AZT, became the first drug approved to treat AIDS (Fitzsimons, 2018). In modern society, more drugs have been developed to treat AIDS and are often used together in methods commonly known as antiretroviral therapy (ART) or highly active antiretroviral treatment (HAART) (History, 2021). AIDs related deaths began to decline in 1995 due to the development of more treatment medications and the introduction of the HAART method (History, 2021). While this decline in cases sounds promising, from 1993 to present day, AIDS is a leading cause of death in men who are aged 25-44 and the third leading cause of death in women (Oldstone, 2020, p. 299).

Ebola

The first cases of hemorrhagic fever (caused by viruses of the Filoviridae family) occurred at around the same time in both South Sudan and the Democratic Republic of the Congo in 1976, one named the Sudan virus and the other the Ebola virus as they were identified as different strains (Malvy et al., 2019). The term Ebola came from the river of the same name in Zaire (the now Demcratic Republic of the Congo), which was close to the outbreak (Malvy et al., 2019). Of the 318 individuals who became infected with the Ebola virus in Zaire 280 died, which is a mortality rate of 88% (Oldstone, 2020, p.223). Since 1976, there have been more than 20 known outbreaks of Ebola in Sub-Saharan Africa, mainly localized to Sudan, Uganda, Democratic Republic of Congo, and Gabon (Malvy et al., 2019). Two other strains have been identified that can infect humans: the first being the Taï forest virus which infected a researcher who had performed a necropsy on a chimpanzee in South Western Côte d'Ivoire in 1994 and the second strain named the Bundibugyo virus that was identified in an outbreak that occurred in Uganda in 2007 (Malvy et al., 2019). Some of the signs and symptoms of Ebola include diarrhea, fever, and lethargy, but unfortunately, these clinical profiles can often be confused with other ailments such as the previously discussed yellow fever (Oldstone, 2020, p.227). The largest Ebola outbreak to date occurred from 2013-2016 (Malvy et al., 2019). This outbreak was centralized to West

Africa, mainly affecting the countries of Guinea, Sierra Leone, and Liberia (Malvy et al., 2019). In this epidemic, 28, 610 Ebola cases were identified though the true amount of cases may have been higher due to under-reporting (Malvy et al., 2019). Of the Ebola cases that were reported, over 11,000 deaths occured (Malvy et al., 2019). While past treatment plans during outbreaks have included only supportive care (ie. rehydration) and treating specific symptoms to improve survival, two monoclonal antibodies known as Inmazeb and Ebanga were approved to treat Zaire Ebola virus by the US Food and Drug Administration in late 2020 (World Health Organization, 2021).

SARS (severe acute respiratory syndrome) and COVID-19

SARS was the first new viral pandemic of the 21st century (Oldstone, 2020, p. 247). The pandemic began mysteriously in southern China in November of 2002 and ending in 2004 (Oldstone, 2020, p. 247). At first, the Chinese government denied that SARS existed until whistle blower Dr. Jiang Yanyong caused them to reverse their stance (Oldstone, 2020, p. 248). This infection spread to 33 countries across 5 continents infecting more than 8000 individuals and killing 774 of them (Oldstone, 2020, p. 247). It was determined that SARS was caused by a novel coronavirus in March of 2003 (Cherry, 2004). The term 'corona' refers to the crown-like shape of the coronavirus (Oldstone, 2020, p. 248). Symptoms of SARS include fever, cough, shortness of breath, and difficulty breathing (Cherry, 2004). The pandemic was ended by SARS perplexingly disappearing, just as mysteriously as it came to be (Oldstone, 2020, p. 247).

In December of 2019, a new coronavirus now named COVID-19 that shares many similarities to the SARS virus appeared in Wuhan and quickly spread through China and the rest of the world, resulting in a global pandemic (Zhu et al., 2020). By March 2020 alone 4,291 deaths occurred across 114 countries

(Sullivan, 2020). While the virus pandemic alone was a big challenge, it was followed alongside what Zhu et al. calls an 'infodemic', which describes the mass spread of false information, resulting in mass global panic. At the time of writing this article, vaccines are being produced to help protect individuals from this virus, including the Pfizer and Moderna vaccines (Campos-Outcalt, 2021). The COVID-19 pandemic is an on-going piece of virus history.

Conclusion

In conclusion, many viruses have existed across history that have significantly impacted various regions across the globe. Different knowledge about viruses was acquired during each infection or pandemic, and many scientists could make breakthrough discoveries or developments. This history of viruses is continuing to evolve with new viruses and mutations, including the COVID-19 virus. As years go by, new viruses, techniques, and scientists will continue to make their mark on the history of virology.

HOW WERE VIRUSES DISCOVERED?

YI YANG FEI

On August 13th, 1865, a man by the name of Ignaz Semmelweis was left to die alone from sepsis in an asylum for the mentally ill in Vienna (Schreiner, 2020). Two weeks prior to that, he was abused and openly wounded when trying to escape from where he initially thought was just a place to relax (Schreiner, 2020). Semmelweis was already suffering from burnout and diabetes, and the opposition from the scientific community, along with the shame from his friends and family that brought him there, added to the emotional toll in his final days (Schreiner, 2020). Even though he passed away in despair, he stood firmly with his last breath on the importance of hygiene against puerperal fever, which is known today as maternal sepsis (Pittet & Allegranzi, 2018; Schreiner, 2020).

Such was the death of Dr. Ignaz Semmelweis, a man later known as the "father of infection control," "father of hand hygiene," and "savior of mothers" (Tyagi & Barwal, 2020). Some portrayed him as a victim of personal tragedies, such as Willy Miksch (1967, as cited in Schreiner, 2020) who described Semmelweis' loneliness and pain in detail for a heroic character in a children's book, or Louis-Ferdinand Céline (1999, as cited in Pittet & Allegranzi, 2018) who thought Semmelweis' wisdom was unmatched against the significance of his findings. Others found Semmelweis' character flaws to be the cause of his tragic fate (Nuland et al., 2006, as cited in Schreiner, 2020). Opal (2010) and Best & Neuhauser (2004) further logically deduced both personal and societal influences behind the rejection that Semmelweis received from scientific communities. Regardless of how his name is known to scholars today, the significance of his findings is nevertheless still relevant (Tyagi & Barwal, 2020). One hundred fifty-five years after his death, on March 11th, 2020, the World Health Organization (WHO) announced the coronavirus disease 2019 (COVID-19) as a global pandemic (Liu et al., 2020). As a result of the worldwide impact of COVID-19, the global awareness on hand hygiene as a form of disease transmission prevention has developed immensely (Rundle et al., 2020). Semmelweis' recommendation and guidelines for hand scrubbing was the foundation of present-day WHO concepts of sepsis prevention and hand hygiene campaigns (Pittet & Allegranzi, 2018).

Science involves increments of revolutionary changes that challenge pre-existing dogma with novel findings or techniques (Artenstein, 2012). Despite the relevance of his findings to the current hygiene guidelines, Semmelweis' efforts were only retrospectively made significant after two decades following breakthroughs in the scientific community (Best & Neuhauser, 2004). In particular, the emphasis of hand hygiene was strongly influenced by findings of Pasteur, Koch, and Lister involving the germ theory and antiseptic techniques (Best & Neuhauser, 2004). The accepted standards of hygiene in healthcare today would have been considered unacceptable by scholars 155 years ago, which could be predicted from the antagonistic attitude against Semmelweis' advocacy for hand-washing (Best & Neuhauser, 2004). Similarly, greek philosophers would have been shocked to know that diseases are caused not by sins but microbes, which are organisms that are invisible to the naked eye but can be viewed in an intricate mirror device known as the microscope (Opal, 2010). Correct thinking may have been limited by society or techniques at the time, but the efforts of pioneers will not be futile. Past principles were undeniably integral in promoting novel findings that either support or refute these ideas to allow subsequent scientific explorations to continue (Artenstein, 2012).

Nevertheless, these old concepts can be epistemological barriers to acceptance of new knowledge, so it is important to not disregard any novel findings that are inconsistent with existing dogma (Artenstein, 2012). The discovery of what is known today as "virus" came through episodes of revolutionary progress, and new findings will inevitably continue to modify and rediscover its definition (Summers, 2014; van Helvoort, 1994). As described by van Helvoort (1994), the construction on the modern concept of virus in the 1950s was accomplished through the deconstruction of the filterable virus concept derived in the past decades. Because of the emphasis on common pathological effects on viruses, however, some degree of continuity between the two concepts was maintained surrounding the effect of viruses on human diseases (van Helvoort, 1994). This chapter will outline the historical discovery of what is known today as a virus. By observing the properties of viruses through the lens of past scientists and their experiments, it is hoped that the modern definition of virus would be appreciated as a fluid concept subject to future deconstructions upon encountering new findings that refute past assumptions (Artenstein, 2012; van Helvoort, 1994).

Development of the Germ Theory of Disease

Infectious diseases were an integral part of the development of human immunity and were present since the beginning of civilization (Opal, 2010). Upon adapting to a civilized society from hunter-gatherer lifestyles, humans were exposed to dense populations with unsanitary living conditions surrounded by decomposing waste (Opal, 2010). Furthermore, domestication of animals and plants led to epizootic infectious agents eventually crossing species barriers to cause epidemics in humans (Opal, 2010). In such environments came the religious explanations for diseases involving punishments or evil spirits which were thought to require magic or exorcisms to cure (Karamanou et al., 2012; Opal, 2010). It was not until the 6th century BC that philosophers like Hippocrates provided a scientific approach to explaining illnesses (Kannadan, 2018). In particular, Hippocrates suggested that diseases are caused by poisonous air in the environment, which led to the development of the miasma theory of diseases (Kannadan, 2018; Karamanou et al., 2012). It was believed that miasma, which were known to be smelly, poisonous substances emitted by decaying organisms, mold, or dust particles, would enter and harm the patient's body (Kannadan, 2018; Karamanou et al., 2012). As a result, sanitation guidelines began to develop as people made efforts to avoid foul-smelling air or other environmental factors in epidemics rather than focusing on individual infections (Kannadan, 2018; Tognotti, 2011). Although these preventative measures were effective to a certain extent, this theory of disease transmission failed to prevent the cholera epidemic in the mid-19th century (Kannadan, 2018).

In 1546, when the majority still believed firmly in the miasma theory of disease, an Italian physician named Girolamo Fracastoro proposed a new theory in his work "De Contagione" (Opal, 2010; Pesapane et al., 2015). From observations and knowledge of past epidemics, Fracastoro proposed the first scientific concept of the germ theory of disease: the idea of tiny invisible particles acting as seeds of infectious diseases by being transmitted between people through direct or indirect contact (Opal,

2010; Pesapane et al., 2015). At the time, however, Fracastoro saw germs as non-living chemical particles that are susceptible to evaporation and diffusion, and that diseases require binding of these seeds to specific tissues in the host to elicit chemical reactions (Brock, 1999, as cited in Karamanou et al., 2012). Nevertheless, Fracastoro's proposal is distinct and significant for its time, when both scientific reasoning and techniques were relatively underdeveloped (Karamanou et al., 2012).

In the subsequent couple of centuries, although little changed in the public stance on the miasma theory, various scientific findings and proposals have paved the way toward the acceptance of this new theory of disease causation (Kokayeff, 2012; Opal, 2010). In 1676, Antonie van Leeuwenhoek, a Dutch textile merchant and also a self-taught scientist, crafted his own single-lensed microscopes and used them to observe small organisms that he referred to as 'animalcules' (Artenstein, 2012; Gest, 2004; Opal, 2010). This marked the first discovery of bacteria through microscopy, occurring about a decade after the first discovery of microscopic fungi by Robert Hooke in 1665 (Gest, 2004). As the concept of microbiology began to develop, the Viennese physician Marcus von Plenciz expanded on the germ theory in 1762 by suggesting the possible living characteristics of germs rather than them being non-living particles (Walker, 1925). Von Plenciz also proposed that each infectious disease has its unique agent (Walker, 1925). Then, in 1840 came the essay "On Miasmata and Contagio" by the German physician and anatomist Jakob Henle (Krieger, 1992). Building off of the previous speculations, Henle proposed that contagions are living materials that are produced and secreted by the bodies of sick individuals, which can then be directly or indirectly transmitted to healthy individuals to cause diseases (Howard-Jones, 1977). In particular, he highlighted a parasitic relationship between the contagions and the sick body rather than the contagion originating from it (Howard-Jones, 1977). Regardless of the retrospective significance of these concepts to the modern germ theory of disease, however, they were still unvalidated theories due to limited scientific techniques at the time and were not widely accepted by the medical and scientific communities (Daniel-Ribeiro & Lima, 2020; Opal, 2010).

A major support for the germ theory of disease appeared when cholera, an infectious disease characterized by explosive diarrhea, dehydration, and death, arrived in England in 1831 (Goldstein, 2011). For the majority who still believed in the miasma theory of disease, the random outbreaks and high fatality made cholera a mysterious and frightening disease (Tognotti, 2011). From their point of view, cholera is caused by direct exposure to decaying matter and is not contagious, since physicians remained healthy even if they frequently treated cholera patients in close contact (Tognotti, 2011). However, precautions to avoid filthy air were ineffective in reducing cholera onset (Kannadan, 2018). In 1854, John Snow, a London physician known today to be the father of epidemiology, challenged the miasma theory through demonstrating the waterborne nature behind cholera using interviews and statistical assessments (Goldstein, 2011; Opal, 2010). As his most influential recommendation, the Broad Street pump handle was removed due to concerns over contaminated water being the source of cholera transmission (Opal, 2010). This effectively halted the cholera outbreak with the support of other sanitation suggestions such as boiling water, washing clothing of the sick, and isolation of diseased from healthy residents (Opal, 2010). Even though the etiological agent in the cholera outbreak, Vibrio cholerae, was not isolated until 30 years later by Robert Koch in 1884, Snow was able to implement public health measures to prevent future outbreaks through epidemiological study alone (Opal, 2010; Tognotti, 2011).

The last necessary puzzle for the germ theory was Louis Pasteur's work on fermentation of liquids in 1857 (Opal, 2010).

Along the same lines of the miasma theory of disease, the theory of spontaneous generation was widely supported by the majority at the time (Karamanou et al., 2012). This theory can be dated back to Aristotle's work in 4th century BC that suggested the existence of a vital force in non-living things to eventually give rise to living things (Lennox, 2001, as cited in Karamanou et al., 2012). In 1668, the Italian physician Francesco Redi designed a series of experiments to disprove the spontaneous regeneration theory: he put meat into different jars with lids that had varying degrees of air flow, then observed growth of maggots from them (Karamanou et al., 2012). As no maggots were able to grow in containers with either closed or netted jars, Redi concluded that living organisms are unable to arise from non-living objects (Parke, 2014). However, Redi incorrectly theorized that galls, which are abnormal growths on plants due to parasitic insects, gave rise to the insects on their own, and this viewpoint continued to act as support for the spontaneous generation theory (Parke, 2014). Louis Pasteur's experiment in 1857, however, strongly supported the germ theory of disease and led to the technique of pasteurization in later eras (Opal, 2010). The French chemist designed a flask with an S-shaped neck and boiled meat broth in the flask (Karamanou et al., 2012). Since airborne microorganisms were unable to ascend the neck of the flask, they settled by gravity and were unable to grow in the broth (Karamanou et al., 2012). When the flask is tilted to allow some broth to enter the bottom of the S-shaped neck, however, the broth became cloudy as microorganisms settled at the bottom (Karamanou et al., 2012). This directly refuted the possibility of miasmatic air causing microbial growth while also demonstrating the effectiveness of sterilization in controlling microbial growth (Opal, 2010). A decade later in 1867, British surgeon Joseph Lister applied the idea of chemical sterilization to surgeries and used disinfectants to prevent infections from open wounds (Opal, 2010). At this time, because of the establishment of proofs for the germ theory of disease, Lister's intervention towards antisepsis widely received acceptance in contrast to Semmelweis' failure in convincing those who believed in the miasma theory (Opal, 2010).

Filterable Agents: An Exception to Koch's Postulates

Louis Pasteur's influence continued to spread as the Pasteur Institute became an international center for various areas of science including microbiology, immunology, and medicine (Opal, 2010). In another part of the world, a Prussian physician named Robert Koch began his career in microbiology and would soon become a contemporary microbiologist as well as a scientific rival to Pasteur (Opal, 2010). In 1876, while performing inoculation studies for the anthrax disease as an appointed district medical officer, Koch discovered the anthrax bacillus and was able to observe bacterial growth using his own oil immersion lens (Blevins & Bronze, 2010). He found that anthrax rods could be propagated for generations between animals and are required for the development of anthrax, which makes him the first scientist to link a specific bacterium to a disease (Blevins & Bronze, 2010). In subsequent years, he continued to expand the field of microbiology through efforts on light microscopy and culturing techniques, including the procedures for isolation of bacterial culture on solid media (Blevins & Bronze, 2010; Opal, 2010). After developing new stain techniques, Koch discovered the tubercle bacillus, the causative agent in tuberculosis, in 1882 (Blevins & Bronze, 2010; Opal, 2010). Following his steps, his assistant Friedrich Loeffler and pupil George Gaffky found glanders, diphtheria, and typhoid bacillus in the subsequent years (Blevins & Bronze, 2010). In 1883, as advancements in microscopy techniques opened more opportunities for microbiology research, Koch, with the support of his assistant Friedrich, established a set of criteria that must be met for a bacteria to be considered the causative agent for a particular disease (Artenstein, 2012; Opal, 2010). Koch's postulates included the following statements: the pathogen must account for clinical and pathological features of the disease in every case of occurrence; the pathogen must not be found as a non-pathogen in other diseases; once isolated and cultured, the pathogen is able to induce the same disease in an animal model; and the pathogen must be able to be re-isolated from the original experimental host (Artenstein, 2012; Opal, 2010). These guidelines influenced an era of etiological discovery of bacterial agents for diseases and are still valid to a certain extent in modern microbiology, despite the limitations on application due to the varied effects on different organisms (Artenstein, 2012; Blevins & Bronze, 2010; Opal, 2010).

Around the same time, another invention brought significance to microbiology research: the Chamberland filter (Summers, 2014). Charles Chamberland, who was an associate working with Louis Pasteur, noted the difficulty to isolate microbes using standard paper filters (Summers, 2014). As the needs for sterilization increased in both research and industrial fermentation processes, Chamberland introduced a type of unglazed terra cotta which contained pores that were permeable to water but retained bacteria (Summers, 2014). Once the filtration is completed, the filter could be sterilized by heat or steam to destroy organic matter that clogged the filter pores (Summers, 2014). The Chamberland filter was commonly used for producing bacteria-free water and is still in use today (Summers, 2014). However, this filter was significant for another reason: the eventual realization of the filterable infectious agents for a botanical disease that were in reality not bacteria at all (Summers, 2014).

Just as Koch's postulates became the standard guidelines in microbiology research for disease etiology, Adolf Mayer, a German agricultural chemist, encountered an unexpected result when filtering infectious agents for his experiments (Artenstein,

2012). Since 1879, Mayer's laboratory in the Netherlands has focused on a disease on tobacco plants presented with pigmented spots on leaves, which he named the "tobacco mosaic disease" (Artenstein, 2012; Bos, 1999). Although Mayer was unable to isolate the causative agent in diseased plants, he was able to inoculate healthy plants with sap from diseased plants to transmit the disease (Artenstein, 2012). Notably, he announced that the agent was no longer infectious after passage of the sap through two layers of filter paper (Artenstein, 2012; Bos, 1999). He reported this finding in 1882 and hypothesized the contagium for tobacco mosaic disease to be soluble and enzyme-like (Artenstein, 2012). Eventually, he continued to believe that these infectious agents were bacterial, even though he was unable to isolate or view them under a microscope (Bos, 1999). At the time, his colleague and co-founder of the Natural Science Society, Martinus Beijerinck, repeated Mayer's experiments to arrive at the same conclusion and also failed to demonstrate the presence of bacteria that could account for the observed effects of disease transmission (Artenstein, 2012; Bos, 1999). Soon, he left academics to work in the field of industrial microbiology (Artenstein & Artenstein, 2010).

A decade later in 1892, Dmitri Ivanovsky, a student studying botanical science in Russia, repeated Mayer's experiments and confirmed the ability of disease transmission (Artenstein, 2012; Bos, 1999). However, he refuted Mayer's claim and stated that the filtrate remains infectious after the sap was passed through the Chamberland filter (Bos, 1999). Due to the heavy influence of Koch's postulates, Ivanovsky had no doubts that the causative agents should have been bacteria and that his irregular findings must be due to equipment defects or laboratory errors, since the Chamberland filter should have prevented the filtration of bacteria (Bos, 1999). Even after the identity of the contagium was eventually revealed, Ivanvsky continued to apply bacterial characteristics to his past findings and failed to recognize the significance of his discovery (Bos, 1999). Around the same time, Beijerinck, who had made great contributions to the industrial microbiology in yeast production, returned to academia as a professor and chair of bacteriology at the Technical University of Delft in 1895 (Bos, 1999). Beijerinck picked up his prior inquiry into the tobacco mosaic disease and observed the same results as Ivanovsky without knowing his presence until much later (Artenstein & Artenstein, 2010). He used the term contagium vivum fluidum, or a soluble living germ, to describe the mysterious nature of the causative agent that was filterable through the porcelain Chamberland filters (Artenstein & Artenstein, 2010). This observation was replicated using solid agar media, where the infectious agent was shown to be nonparticulate and soluble, yet still induced disease in serial transfers (Summers, 2014). Furthermore, Beijerinck demonstrated the inability of this agent to grow independently where living and dividing host cells are not present (Artenstein, 2012; Beijerinck, 1898, as cited in Artenstein & Artenstein, 2010). Along with his proposed experiments, Beijerinck's description of this new class of infectious agent as the "filterable virus" marked the beginning of the field of virology and subsequently the golden age of vaccinology for infectious disease prevention (Artenstein, 2012; Summers, 2014).

The Arrival of Modern Virology Concepts

With Beijerinck's findings came waves of similar filterable agents reported in literature that were previously considered as incompatible with Koch's postulates for bacteria (Summers, 2014). Within three decades after the concept of filterable infectious agents was established, at least two dozen diseases were found to be associated with filterable viruses (Helvoort, 1996, as cited in Claverie & Abergel, 2016). Nevertheless, this finding was not a prerequisite to successful intervention to prevent viral disease transmission, as Edward Jenner had established the procedures for smallpox vaccination a century ago in 1796 using matters from cowpox lesions (Riedel, 2005). However, similar to the impact of Koch's postulates, the scientific community used Beijerinck's observations as strict guidelines to categorize a long list of diseases with unknown origins involving filterable agents (Summers, 2014). Thomas Rivers (1932), an American virologist, described two prevailing sets of criteria to categorize viruses from other pathogenic microorganisms (Helvoort, 1994). The positive characteristic of the filterable virus was the ability of its filtrate to multiply and infect host organisms (Helvoort, 1994). The negative characteristics, which were greatly influenced by Beijerinck's findings, involved invisibility under microscopy, inability to be retained by laboratory filters, and inability to propagate when living host cells are absent (Rivers, 1932).

Evidently, there was a greater emphasis on distinguishing filterable viruses from bacteria rather than focusing on their qualitative characteristics as infectious agents (Summers, 2014). Consequently, the scientific community focused on quantitative analysis of virulence using serial dilutions (Summers, 2014). This changed when the bacteriophage was discovered in 1915 and 1917 by two independent researchers: Frederick Twort and Felix d'Herelle (Taylor, 2014). Distinct from the other filterable viruses at the time, these were able to attack bacteria colonies directly, which earned them the name "ultra-microbes" (Summers, 2014). This conclusion came from the observation of contaminated areas being unculturable on an agar culture plate, which was described as transparent and glassy-looking by Twort (Summers, 2014; Taylor, 2014). D'Herelle made the same observation and hypothesized that these filterable viruses were able to infect bacteria, thus giving them the name to reflect on the bacteria-eating nature (Taylor, 2014). However, he also found plaque growth to be proportional to the dilution of the phage solution, which contrasted against the common belief of viruses being soluble in the infectious filtrate (Summers, 2014; Taylor, 2014). In other words, phages were corpuscular particulates and not soluble enzymes (Summers, 2014). In 1935, Wendell Stanley was able to crystallize the tobacco mosaic disease virus (Claverie & Abergel, 2016). Four years later, the physical appearance of phages were observed by electron microscopy in 1939 by Kausche et al. (1939, as cited in Summers, 2014). In the same year, Ellis & Delbrück (1939) recognized the presence of a latent period that was present in bacteriophages but not bacteria, which immediately undergo binary fission. By this point in time, virus was widely accepted as its unique entity distinct from bacteria (Claverie & Abergel, 2016).

Conclusion

The development of the field of virology was supported by centuries of work on microbes and infectious diseases, and it would be impossible to attribute one single finding as the most significant event that built up to the acceptance of various theories in virology. However, in no way is this the end of discoveries. The germ theory of disease has already been challenged as recent researches in the field of microbiology suggest an importance in gut dysbiosis in disease causation rather than a single pathogen being the causative agent (Cabrera-Perez et al., 2017). In 1992, initiated by Tim Rowbotham's recognition of the mimivirus, a family of giant viruses were discovered that were visible under the microscope and filterable by the Chamberland filter (Claverie & Abergel, 2016). In contrast to those of most viruses, genomes for giant viruses coded for not only functional and structural proteins for hijacking the cell machinery but also its own organelles to become the site of translation, transcription, and replication (Claverie & Abergel, 2016). As of 2016, there were already four distinct families of giant viruses whose discoveries were historically hindered by strict adherence to past principles of size and appearances (Claverie & Abergel, 2016). The stories of Semmelweis, Snow, and many others with retrospectively significant findings illustrate the importance of acknowledging findings that could not be explained by the prevailing dogma, such as Koch's postulates at the time (Artenstein, 2012). After all, scientific advancements are only made possible through revision and rigorous observations that overcome epistemological obstacles to knowledge (Artenstein, 2012).

WHAT IS THE IMPACT OF VIRUSES? - MACRO SOCIETAL VIEW

TIM CHAPMAN

Throughout history viruses have come like waves, being brought into a community, ravaging throughout before moving on to the next in search of more viable hosts. Scientific advancements made throughout the 17th and 18th centuries have given humanity the fighting chance against viruses. Discoveries such as Antonie van Leeuwenhoek first discovering bacteria, calling them "little monsters" in his papers, or John Snow who despite the odd's persuaded Queen Victoria to remove the Broadwick Street water pump handle, nearly ending the communities devastating cholera outbreak, while discovering that Cholera was a water-borne virus (Lane, 2015, Snow, 1849). These discoveries focused on the medical side of the virus, discovering what it is and how it spreads,later discoveries included how to fight back against viruses. Despite the advancements made in the field of medicine regarding viruses, when a virus is allowed to spread quickly enough, every aspect of society affected. The impact of a virus is felt far beyond the steps of any hospital or research centre, when a virus reaches the level of a pandemic, the basic aspects of society become interrupted. Throughout history, the impact of a virus are primarily the same across the board, with globalization being a major factor affecting the impact that viruses have had in modern times. This chapter takes a macro approach to how a virus affects society, unearthing how a pandemic affects the global supply chain, how easily hospitals can become overwhelmed, the economic impact of viruses as well as the positive impact that viruses can have on society.

The scene of empty grocery store shelves, streets packed with abandoned fuelless vehicles, and overflowed hospitals are all common in Hollywood portrayals of life during a pandemic but are seen as unrealistic or unimaginable. The reality is that viruses have the potential to be extremely impactful towards the society's supply chain if the spread has reached a specific level or area of the world. Depending on the specific virus outbreak that you research, the effects on the supply chain will vary. This is because the scale of the supply chain will vary depending on the historical period the virus occurred, for example, during the bubonic plague citizens did not have to worry about the local grocery store running out of milk. Another aspect to consider with regards to the sheer scale that is in place for our modern supply chain to function in the capacity that it does. If a manufacturer has to close a plant due to a local outbreak of a virus, then the product made by the facility will not be produced, as an interactive example, let's say that product is ball bearings. The local ball-bearing plant is closed for a mandatory quarantine of an undisclosed amount of time (as different viruses require different quarantine times), the next factory in the chain requires ball bearings to build wheel bearings for semi-trucks, now the local semi-trucks are limited on the availability of replacement wheel bearings, if there was little stock, to begin with, now there is a crisis.

This may seem like an extreme example, but every product that is manufactured has a purpose somewhere, even if one small piece is missing, it has the potential to have major consequences further down the supply chain. The supply chain has had several challenges throughout its history and through this, it has become very resilient. Supply chain resilience can be defined as: "the capacity of a supply chain to persist, adapt, or transform in the face of change" (Wieland & Durach, 2021). During the early stages of the Coronavirus pandemic (COVID-19), the world witnessed what occurs when the supply chain lacks raw material that is needed to manufacture another product, for example toilet paper. This reduction in raw materials and high demand due to stockpiling, caused by the COVID-19 pandemic, resulted in mass shortages across the globe for a short period (Paul &Chowdbury, 2020).

With an understanding of how huge an impact a minor outbreak of a virus could have on the supply chain, let's look at the effects the outbreak of a virus could have on a hospital setting. Beginning with the obvious, during an outbreak the hospital will be the first place that will be overrun with sick people. In modern times, the hospital is a central institution within society, once the hospital beds are full there is nowhere else to go, the only option is to create new areas for the sick to occupy. The next obvious point to make is with regards to staffing, once the virus makes its way to the hospital all staff are at increased risk compared to the general population, with the risk of catching an illness varying by position (Jecker, 2020).

A study completed during the 2009 H1N1 pandemic found that staff in their younger years were found to have higher anxiety surrounding the virus than staff members who were older (Matsuishi et al., 2012). The same study also found that front-line nurses had higher levels of anxiety than the doctors who would see the patient after someone had made sure they were as safe as possible (Matsuishi et al., 2012). Outside of the limitation surrounding the number of hospital beds and staff concerns, is the lack of personal protective equipment (PPE) that arises during high levels of stress such as a pandemic. During times of high stress on the hospital system, acquiring PPE can look like this: "Deals, some bizarre and convoluted, and many involving large sums of money, have dissolved at the last minute when we were outbid or outmuscled, sometimes by the federal government. Then we got lucky, but getting the supplies was not easy" (Artenstein, 2020). The difficulty in acquiring PPE during an emergency compounds the already difficult situation to manage. Taking an even more macro approach, the consequences experienced by a weakened supply chain make up only a portion of the consequences experienced by the economy as a whole.

The supply chain makes up a portion of the economy and from the previous paragraph, it is clear that a small disruption can have grave effects, let's take this logic and apply it to the macro view of the entire economy. When a virus reaches the level of a pandemic it has already spread between international borders, with this means the travel, tourism and retail sales and all other associated businesses are compromised (Wishnick, 2010). According to the World Health Organization (WHO), from November 2002 to July 2003 Severe Acute Respiratory Syndrome (SARS) was rampant in China, and across the world (Wong & Leung, 2007). Following the 2003 SARS outbreak, financial analysts have concluded that the virus cost the Chinese government one percent of its national gross domestic product, and an estimated total figure of anywhere between thirty and one-hundred billion USD (Smith, 2006, MacKellar, 2007).

With areas of the economy such as tourism and travel being reduced, this not only affects the large corporations and governments but is also linked to an increase in the unemployment rate. During the SARS outbreak the Chinese recorded an eight percent drop in its migrant worker population, accounting for roughly eight million people, in addition to this, roughly fifteen million Chinese citizens lost their jobs when the government closed the service industry in favour of stricter public health orders (Lee & Warner, 2006). With a significant reduction in the labour force during any viral outbreak, the government needs to play an essential role in the prevention and control plans that work to support the economy.

An important aspect of maintaining order in the event of a virus outbreak is the ability of the local/national government's approach the situation. The main responsibilities of governmental roles during a large-scale virus are the prevention and control plans and the ability to reduce panic. During the beginning phases of SARS, there was a lack of trustworthy official information, this resulted in many folk tales and rumours being spread that was not scientifically backed, further prompting the panicked public to mass buy counterfeit remedies (Ma, 2008). In a later viral epidemic, the Chinese government was able to learn from the past and control panic more effectively.

In February 2013, China confirmed their first case of H7N9 avian influenza, a virus with a higher fatality rate (Liu et al., 2013). "The government management of the health and agriculture sectors was completely open" and in turn calmed the citizens, removing the aspect of panic (Lee & Warner, 2006). A later National 12320 Telephone Management Centre opinion survey asking about the government's response to the H7N9 virus, 80 percent of respondents were satisfied with the government's steps towards prevention and control (Lee & Warner, 2006).

It is not only the government's ability to control the public in times of stress but to also guide the private business sector. Industrial mobilization is a term that transforms an industry from its regular manufacturing to supply necessary supporting materials to a cause, most typically military objectives. During World War II, the United States of America called manufacturers to aid in making war materials, producing about forty percent of the world's total munitions (Morgan, 1994). Following SARS, companies such as 3M have made efforts to provide PPE including surgical masks and respirators preventing viruses such H1N1 and Ebola and COVID-19. (MacIntyre et al., 2014, Dzyakanava et al., 2014, 3M, 2021). 3M is not the only company to have stepped out of their regular manufacturing to aid in essential product production. The WHO in the face of COVID-19 declared a global shortage of alcohol-based hand rubs (ABHRs), by relaxing legislations in some countries such as the USA, Australia and the UK, it was easier for companies such distilleries to manufacture and distribute ABHRs (HM Revenue & Customs, 2020, Cook, J., 2020, U.S Department of Health and Human Services, 2020). Early on in the COVID-19 pandemic, locals in Edmonton, Alberta had volunteered to make and donate masks in preparation for a potential PPE shortage that the USA had already experienced (Weisberg, 2020). From businesses to at-home volunteers, essential resources can be produced to help in the efforts to fight against a large-scale virus.

Zika is a mosquito-borne virus that affects newborns with microcephaly or small head and brain, and pregnant women (Qureshi, 2017). In November 2015, Brazil had declared the epidemic a national health emergency and the WHO declared its high potential to spread across the Americas (Qureshi, 2017). Across Latin American countries Zika has caused potential damages to economic growth. According to the World Bank Group, "Initial estimates of the short-term economic impact of the Zika virus epidemic for 2016 in the Latin American and the Caribbean region (LCR) are a total of US$3.5 billion, or 0.06% of GDP" (Ulansky, 2016). These numbers were calculated using the effects of the crisis on economic staples such as tourism, worker productivity and negative media attention (Qureshi, 2017). Aside from direct expenditure, the loss of productivity was observed in roughly twenty percent of the population with employees missing work up to a week at a time, due to the virus (Zagorsky,

2016). This resulted in financial loss and hardships on small local businesses.

This loss was also felt by the tourism industry in Brazil, a country that relies on such an industry, seeing vacation cancellations upwards of $250,000 in the first half of 2016, while the overall tourist industry felt a loss of 3.5 billion dollars during the overall epidemic (Qureshi, 2017). Negative media also hindered both tourism and worker productivity as individuals avoided work where case counts were high, and tourism saw a quick decline in travel, particularly in pregnant women, as the news broadcast the situation globally (Qureshi, 2017). Where networking originally caused fear in foreign and domestic populations, the current aim is for the radio to inform the public about prevention and protection from the virus and the mosquitoes that carry it (United Nation Educational, Scientific and Culture Organization, n.d.).

It is through increased education that a rise in international collaborations has created positive impacts. Human Immunodeficiency Virus (HIV) is most present in third world countries, specifically in Sub-Saharan Africa where one in twenty-five adults live with the virus (Government of Canada, 2017). "There was a 45% decrease in new infections between 2000 and 2015," this decrease is credited to new drugs and treatments, improved health services, prevention programs and public awareness campaigns (Government of Canada, 2017). Since its involvement in UNAIDS in 1996, the Joint United Nations Programme on HIV/AIDS, Canada has contributed over $100 million towards the cause, the main area of concern being education and providing condoms so that there are no shortages of protection (Government of Canada, 2017). Canada works closely with its UN partners and supports UNAIDS 90-90-90 goal: where 90% of all people living with hiv will know their status and received antiretroviral (ARV) therapy, and of those who have had ARV have viral suppression (Topic: HIV treatment. n.d.). It's with this ambition that Canada has aligned with UNAIDS' ultimate goal of eliminating HIV/AIDS by 2030 (Topic: Fast-Track cities. n.d.). Globally, countries have rallied together to administer ARV to ten million people, distribute 5.3 billion condoms, hold 509 million HIV/AIDS counselling/testing sessions, delivered 3.6 million HIV-negative infants from HIV-positive women, and provide basic care and support to nearly 7.9 million orphans and vulnerable youth (Government of Canada, 2017).

From John Snow's first discovery of viruses to the present day, many milestones have been achieved in developing and optimizing current medical research. From data collected between 2008 and 2013, Elsevier concluded that the most international collaborative field was virology with 120 active participating nations (Fry et al., 2020). Using COVID-19 as an example, the global effort to research the virus was immense. The last large-scale collaboration between medical researchers was observed at the peak of the AIDS epidemic in the 1990s as medical professionals sought to combat the disease (Apuzzo & Kirkpatrick, 2020). From a more modern perspective, this collaborative nature can be seen in research and vaccine development for COVID-19. In a typical research environment, pioneering research can lead to grants, tenure etc, however, as Dr. Ryan Carroll a Harvard Medical professor states, "The ability to work collaboratively, setting aside your personal academic progress, is occurring right now because it's a matter of survival" (Apuzzo & Kirkpatrick, 2020). The average time to develop a vaccine is between ten and fifteen years, the fastest vaccine developed before COVID-19 was the mumps vaccine in 1967, taking four years. It is through international collaboration that many viable vaccines for COVID-19 were able to be developed, tested, approved and administered in less than a year (Solis-Moreira, 2020).

Viruses have the potential to make a huge impact on society, the scientific community has been working for hundreds of years to reduce the medical impact that viruses can present. When the environment is right, a virus can quickly spread out of control, when this occurs the impact felt on society can include supply chain disruptions, increased burden on the healthcare system, disasterous effects on the economy but can also have positive effects such as increased international collaboration. It is critical to research the macro point of view when studying viruses as it allows for the chance to gain an understanding of the scale on which even the smallest viruses can have an impact on society.

WHY IS IT IMPORTANT TO STUDY VIRUSES?

RICO CUECACO

When the word 'virus' is mentioned, many are quick to comment on the harmful effects of a virus and its destructive nature on humans. In the contexts of modern medicine, viruses are known to either cause drastic harm to mass populations or serve beneficial to human health. To discover the lengths that viruses affect humans we need to explore the underlying surrounding the study of viruses. Professionals of medicine need to study and understand what viruses are to formulate solutions and appropriate actions for how society can combat them without putting more lives in danger. In recent history, viruses such as Ebola, Polio, SARS, and H1N1 Influenza had caused tremendous effects on mass populations due to their destructive nature. Amid a pandemic, the study of viruses is an important step for humans to understand what makes a virus so dangerous, the effects on the human body, the science behind vaccinations, and how it can prevent further infections and, as well, imply immunity once administered. To answer the question, why is it important to study viruses? We must first explore what makes a virus so harmful.

To understand what makes viruses so dangerous, we must explore biological science. Viruses vary in complexity. They consist of genetic material, RNA or DNA, surrounded by a coat of protein, lipid (fat), or glycoprotein. Viruses cannot replicate without a host, so they are classified as parasitic (Crosta, 2017). In addition, viruses can cause widespread harm to mass populations if the virus itself can find and replicate itself inside of human cells. A virus exists only to reproduce. When it reproduces, its offspring spread to new cells and new hosts. (Crosta, 2017) If the virus has the right receptor, it either begins to insert its genetic information (their DNA or RNA) into the cell, or it inserts itself into the cell. From here, the virus starts replicating itself and infecting the host body. There are two ways that the virus can replicate itself: Through the lytic cycle or the lysogenic cycle. The Lytic cycle is considered the primary and fastest way for a virus to spread amongst several individuals. During this cycle, the virus's genetic material "hi-jacks" the cell and starts using its resources to create more copies of itself. It replicates itself in different parts (e.g. replicates it's DNA/RNA, replicates its capsid, etc). Once the virus has created many copies of its different parts, the virus reassembles all of its parts. It does so until the entire cell is filled to the brim until it explodes and releases itself back into the host body to infect more cells. (Bozanic, 2020) When a virus spreads, it can pick up some of its host's DNA and take it to another cell or organism. If the virus enters the host's DNA, it can affect the wider genome by moving around a chromosome or to a new chromosome. This can have long-term effects on a person. In humans, it may explain the development of hemophilia and muscular dystrophy. Viruses are widely known for harmful reasons, whether the lasting effects are short-term or long-term. Although viruses do great harm to humans, it's ability to transfer from one living organism to another proves that viruses are able to cause harm to different living species. For example, Some viruses only affect one type of being, say, birds. If a virus that normally affects birds does by chance enter a human, and if it picks up some human DNA, this can produce a new type of virus that may be more likely to affect humans in future. (Crosta, 2017) The seasonal flu virus or 'influenza' is spread across continents by the migration of birds and causes around 650,000 deaths each year. Due to imports, exports and many ways to travel across continents, viruses that were once confined to specific populations and continents have now spread beyond their 'natural' borders - causing epidemics and worldwide pandemics. (Burt, 2020) This is why scientists are concerned about rare viruses that spread from animals to people. The real danger of a virus starts when they come in contact with a cell which they will make their host. Looking at it on a global scale, viruses such as COVID-19 have proven that a virus possesses the ability to affect millions of individuals and willingly disrupts the livelihoods of mass populations. One aspect that signals pandemic potential in a virus is having an RNA, rather than DNA, genome. That's because the process of copying RNA typically doesn't include a proofreader like DNA replication does, and so RNA viruses have higher mutation rates than the DNA variety. "This means they can change and become more adaptable to human infection and human transmission," says Steve Luby, an epidemiologist at Stanford University. (King, 2020) What makes viruses so harmful is their ability to cause widespread on a molecular scale, making it difficult for the body to contain viral cells due to the infectious characteristic of viruses. To further our understandings of viruses, we must look deeper into the effects a virus has on the human body.

To understand how a virus works, it is necessary to think on a very small scale. At such a small scale that the human eye could never see the movements of a virus. Viruses are tiny microbes on the planet, yet they can make a person sick and even kill. In extension to ideas mentioned earlier, a virus can inflict severe illness due to the microorganisms infecting human cells. These microorganisms enter the body through the mouth, eyes, nose, genitals, bites, or open wounds. Moreover, they are transmitted through different routes. Some diseases are spread by direct contact with infected skin, mucous membranes, or body fluids. There is also the possibility of indirect contact, when a person touches an object (door, handle, table), which has the virus on it, when an infected person sneezes, coughs or talks or when the mucous membrane comes into contact with another person. In some other cases, the virus is transmitted through a common vehicle such as contaminated food, water or blood. Finally, there are vectors: rats, snakes, mosquitoes etc., which transmit the virus to humans. (King, 2020) It is important to clarify that when a virus infects a human, it does not always end up in disease. The infection occurs when the virus begins to multiply. And, the disease occurs when many body cells are damaged by the infection, which is also when the symptoms and illness appear. Not all viruses cause diseases. Some are harmful due to the virus entering a host cell and applying pressure on the host cell to make copies of the virus and then the virus is released from the host cell (King 2020). According to Medicine News Today, there are two dramatic effects that viruses have on humans. Firstly, viruses in some cases can cause serious life-threatening conditions such as dehydration or pneumonia, which validates the concerning effects that viruses can have on the human body once infectious cells cling onto a host. And Second, Viruses can cause diseases and cancers. It also causes brain damage on some occasions. For example, measles. If in this case, the human body is unable to create anti-bodies to combat infectious cells, the body will slowly deteriorate alongside the dying cells. When the immune system fails to control the virus, a process called pathogenesis begins. The virus crosses obstacles such as distance, the immune system or mucous membranes to reach different organs. Once it begins to replicate, the person will get sick and his/her organs will be infected. Depending on how severe the symptoms are, the person will have to rest or seek medical help (Burt, 2020).Viruses are like hijackers. They invade living, normal cells and use those cells to multiply and produce other viruses like themselves. This can kill, damage, or change the cells and make a person sick. Different viruses attack certain cells in the body such as the liver, respiratory system, or blood. The most widespread infection caused by a virus is a viral disease. This microorganism causes a wide variety of viral diseases such as the common cold. This viral disease is widely known for its effects on the upper respiratory tract and depending on the severity of the case, the common cold has the potential to inflict long-term effects on breathing and the immune system. Viruses cause the immune system to respond and attack them. This response causes stress and inflammation in the body. The effects of this response often leave people feeling down, fatigued, and sometimes depressed. (Bozanic, 2020) For most viral infections, treatments can only help with symptoms while you wait for your immune system to fight off the virus. Antibiotics do not work for viral infections. There are antiviral medicines to treat some viral infections. Vaccines can help prevent you from getting many viral diseases.

Vaccines are widely known to be the primary course of action to contain and prevent further transmissions of viruses. To further understand the prevention of the spread of viruses, we must explore the science behind vaccines and how it implies immunity once administered. Thanks to scientific research and investment from the most developed countries, progress has been made with regards to understanding how a virus penetrates a host cell and proceeds to spread. In theory, this information should provide scientists with enough to establish virus control procedures, along with creating antivirals, antibodies and vaccines. Unfortunately, the primary challenge they face is the lack of ability to cultivate and grow a virus in a laboratory setting.

(Crosta 2017) Although it is impossible to prevent any virus from infecting humans, through the experience of generations and the help of science, the human body should be able to defend itself from foreign agents. It has been proved that vaccines have saved more lives in Canada than any other medical intervention in the past 50 years. Before we had vaccines, many Canadians died from preventable diseases. Vaccines also prevent diseases that are rarely deadly but can cause pain and long-term health problems. (Walker, 2020) Some historical examples that legitimize the effectiveness of vaccines are shown through cases of the Polio disease and Haemophilus Influenza type B (Hib). In the early 1900s, before the introduction of the polio vaccine, thousands of Canadians were paralyzed or died from polio. Thanks to vaccination, Canada has been polio-free for the last 20 years. (Walker, 2020) As for the Haemophilus Influenza, before the introduction of the Haemophilus influenzae type b (Hib) vaccine in 1988, Hib was the most common cause of bacterial meningitis among children younger than 5 years of age in Canada. Every year about 1500 cases of Hib meningitis occurred in Canada in children under the age of 5. Since the vaccine, Hib infections have almost disappeared in Canada. (Walker, 2020) The integration of vaccinations into modern medicine has proven its effectiveness by allowing our bodies to create immunity against threatening viruses. Vaccines don't just protect the people getting vaccinated; they protect everyone around them too. The more people in a community are vaccinated, the harder it is for a disease to spread. If a person infected with disease comes in contact only with people who are immune, the disease will have little opportunity to spread. The type of protection created when most people are vaccinated is called "herd immunity." It means that many of us are protecting each other, and especially the most vulnerable among us, such as babies who are too young to get vaccinated, people who can't receive certain vaccines for medical reasons, and people who may not adequately respond to immunizations such as seniors who possess poor immune systems. (Sandra, 2020) The purpose of a vaccine is to help your body build a defence system to fight foreign germs that could make you sick. To build up your immune system, the body must be exposed to different germs. When the body is exposed to a certain virus for the first time, it will produce antibodies to fight it. However, it takes time for the body to create antibodies to fend off the viruses, so in return, the body will eventually submit to the virus and ultimately cause the individual to become sick. Once antibodies are produced, they will stay in your body to further protect themselves from future interactions with an infectious disease. Most vaccines are a weakened form of the disease germ injected into your body. This is usually done with a shot in the leg or arm. Due to the weakness of the virus, the body will have an easier time creating a defence system against further diseases. These antibodies will then stay in your body for a long time. In many cases, they stay for the remainder of one's life. If the person is ever exposed to the disease again, the body will fight it off without ever getting the disease. Taking into consideration the effectiveness of vaccines, it has been proven that everyone needs to be vaccinated. There is, in fact, a widely accepted immunization schedule available. They list what vaccines are needed, and at what age they should be given. Most vaccines are given to children. It is recommended they receive 14 different vaccines by their 6th birthday. Some vaccines are combined so they can be given together with fewer shots. Evidence surrounding vaccines establishes how effective vaccines are and why it is encouraged to receive them to protect the vast majority of the population.

With the science of vaccines and their effectiveness, it is important to mention how the study of viruses played a major contribution to the innovative creation of vaccines. The motivation for focusing on a specific virus is often its importance in terms of impact on human interests. Taken from ideas mentioned earlier, the study of viruses is an essential step to discover solutions in the case where a virus reaches pandemic levels. The importance of a virus is not due to the virus itself, but to the hosts, they infect and affect, and many viruses are important because they cause diseases in humans, animals, or crops. They are also important because they are active and abundant in aquatic environments, infect key species, and affect community composition and nutrient flow and thus all aquatic ecosystem services. Hence, we need to know about viruses to understand nature and implement knowledge-based management of our resources (Sandaa, 2019). On a global scale, viruses are bound to any country and environment. The study of these environment-specific viruses could contribute to medical advancements in preventing potential pandemic level viruses. Furthermore, studying different types of viruses from different parts of the world would also deliver educational resources regarding threatening viruses. Viruses are continuously changing, evolving into more dangerous and harmful variants. The importance of studying viruses is simply for the safety of future populations when another pandemic level virus takes over the world.

In conclusion, the importance of studying viruses has led to the distribution of vaccines, the educational value in learning about the nature of viruses, and preventive actions that must be considered when a virus is active and at large. The harmful nature of viruses has affected mass populations for centuries, and it seems that viruses will continue to become a part of human lives. Through scientific discoveries and medical advancements, our understandings of viruses and the appropriate measures to protect ourselves from them will continue to evolve, eventually leading to an educated demographic on viruses. In retrospect, viruses have always been a contributing factor to annual death rates. Everyone needs to look and educate themselves on the destructive nature of viruses to combat and protect themselves from future cases of threatening diseases.

WHAT ARE VIRUSES?

CHITRINI TANDON

Viruses are microscopic organisms that exist everywhere on Earth and are classified as parasitic. Viruses can infect animals, plants, fungi and bacteria, they can cause a disease, infection or sometimes have no noticeable reaction (Crosta, 2017). The same virus can have different effects on different organisms. They vary in complexity and consist of genetic material, RNA or DNA, and are surrounded by a coat of protein, lipid or glycoprotein (Crosta, 2017). Additionally, they cannot replicate on their own, meaning they need a host (Crosta, 2017). Viruses are considered the most abundant biological entity on Earth (Crosta, 2017). Some examples of the diseases caused by viruses are rabies, herpes, and Ebola, and while there is no cure for a virus, vaccinations can help to slow down or stop the spread of a virus (Crosta, 2017). Some viruses may only affect one type of being, for example a virus may only affect birds. If the virus that affects birds enters a human and picks up human DNA this can lead to the production of a new type of virus which may affect humans in the future (Crosta, 2017).

Structure and Types

Viruses exist in almost every ecosystem on Earth and before entering a cell the virus exists in a form called virions and are one-hundredth of the size of a bacterium (Crosta, 2017). Virions consist of two or three distinct parts: genetic material (DNA or RNA), a protein coat called a capsid which will protect the genetic information and a lipid envelope which is sometimes present around the capsid when the virus is outside a cell (Crosta, 2017). Viruses cannot make proteins because they do not contain ribosomes meaning they are completely dependent on their host (Crosta, 2017). After coming in contact with a host cell the virus will inject it's genetic material into the host to then take over the host's functions and reproduce. Viruses also have different shapes and sizes which may be; helical, icosahedral or envelope. The helical virus has a helix shape and an example is the tobacco mosaic virus (Crosta, 2017). The icosahedral virus is near-spherical and an example would be most animal viruses (Crosta, 2017). The envelope shape means that the virus has a cover with a modified section of cell membrane which creates the lipid envelope and examples include the influenza virus and HIV (Crosta, 2017). There are also nonstandard shapes which combine both helical and icosahedral. When the virus interacts with the host DNA it causes the virus to change (Crosta, 2017). There are also some good viruses such as Escherichia coli (E. coli) (Crosta, 2017). The simplest virus has enough RNA or DNA to produce four proteins while the most complex ones can encode 100 to 200 proteins (Lodish et al., 2000). All viruses use ribosomes, tRNAs and translation factors for synthesizing their proteins (Lodish et al., 2000). Viruses will take over the cellular machinery for macromolecule synthesis during the later stage of infection. The cellular machinery will then make viral mRNAs and proteins instead of the normal cellular macromolecules (Lodish et al., 2000). An example would be an animal cell infected influenza or vesicular stomatitis virus will only synthesize only one or two types of glycoproteins, which are encoded by virtual genes, and on the other hand uninfected cells will produce hundreds of glycoproteins (Lodish et al., 2000). Most products of the viral protein fall into three categories: special enzymes which are needed for viral replication, inhibitory factors which will stop the host-cell's DNA, RNA, and protein synthesis and lastly, the structural proteins used in the construction of new virions (Lodish et al., 2000). The latter two are made in larger amounts than the first type. There are some viruses which have a genome composed of RNA instead of DNA and they can grow easily and in large quantities because their RNA genome also acts as their mRNA (Lodish et al., 2000).

The protein coat which is called a capsid encloses the nucleic acid of a virion and is composed of multiple copies of one protein or a few different types, each of the copies are encoded by a single viral gene. Due to this reason the virus is able to encode all of their information for making a large capsid in a small amount of genes, this technique is important because only a small amount of RNA or DNA and a limited number of genes can fit into a virion capsid (Lodish et al., 2000). Together the nucleic acid and the capsid are called a nucleocapsid (Lodish et al., 2000). There are two different ways to arrange multiple capsid protein subunits and the genome into the nucleocapsid. The simpler structure is a protein helix with the RNA or DNA protected inside, an example of this helical nucleocapsid is the Tobacco mosaic virus (TMV) (Lodish et al., 2000). In TMV the protein subunits form into broken disk-like structures which resemble lock washers and will form the helical shell of a long rod-like virus when they are stacked together (Lodish et al., 2000). The other major type of virus is called icosahedral or quasispherical virus, it is based on the icosahedron which is a solid object built of 20 identical faces and has an equilateral triangle (Lodish et al., 2000). The simplest of this type has 20 triangular faces which are constructed of three identical capsid protein subunits, meaning there are a total of 60 subunits per capsid (Lodish et al., 2000). All of the protein subunits are in equivalent contact with one another. An example of this type of virus is the Tobacco satellite necrosis virus (Lodish et al., 2000). Simple viruses such as TMV will undergo self-assembly if they are mixed into a solution but more complex viruses will assemble in multiple stages inside a cell.

Animal viruses are a specific type of virus and come in many different shapes, sizes and genetic strategies (Lodish et al., 2000). Some examples include: poliovirus and human immunodeficiency virus (HIV). While there are many different kinds of viruses, most of them will produce the same symptoms or disease states such as runny nose, sneezing or red eyes (Lodish et al., 2000). There are six classes of animal viruses which can be named and it is important to note that bacteriophages and plant viruses are classified in a similar way. Class I and II of viruses consist of DNA (Lodish et al., 2000). Class I viruses contain a single molecule of double-stranded DNA called dsDNA and some examples of this type include; Adenoviruses, SV40, Herpesviruses, and Human papillomaviruses (HPVs) (Lodish et al., 2000). Another type of class I viruses is known as poxviruses, they replicate in the host-cell's cytoplasm and an example is variola which causes smallpox (Lodish et al., 2000). Class II viruses are known as parvoviruses (which come from the Latin word parvo, meaning poor) (Lodish et al., 2000). These viruses are simple and contain one molecule of single stranded DNA or ssDNA. Some of them will enclose both plus and minus strands of DNA (the viral mRNA is the plus and the complementary strand which cannot function as mRNA is known as minus) in separate virions and others will only contain the minus strand (Lodish et al., 2000). In either of the two categories the ssDNA will be copied into dsDNA and then copy itself into mRNA (Lodish et al., 2000).

Class III to VI have RNA genomes and they infect a wide range of animals (Lodish et al., 2000). Class III viruses have double stranded genomic RNA known as dsRNA. The minus RNA strand is the template for the synthesis of the plus strand and typically this class of viruses are known to have genomes of 10 to 12 separate dsRNA molecules (Lodish et al., 2000). Each of these dsRNA molecules will encode for one or two polypeptides and this is why these viruses have "segmented" genomes. The virus itself contains the set of enzymes needed to use the minus strand as a template to make the plus strand after infection (Lodish et al., 2000). Class IV viruses have a single plus strand of RNA which is exactly the same as viral mRNA (Lodish et al., 2000). There are two types of class IV viruses, class IVa which are characterized by poliovirus, viruses which are synthesized from a single mRNA first into a long polypeptide chain which is then cleaved into different proteins. Class IVb synthesizes at least two species of mRNA, one being the same length as the virion's RNA and the other being similar to the 3' third of the genomic RNA. Both are translated into polyproteins (Lodish et al., 2000). Examples of this type of proteins include Sindbis virus and viral encephalitis (Lodish et al., 2000). Class V viruses have a single negative strand of RNA whose sequence is complementary to viral mRNA (Lodish et al., 2000). This strand acts as a template for the synthesis of mRNA. There are two types of these viruses. Class Va viruses include measles and mumps which encode for a single protein from several mRNAs (Lodish et al., 2000). Class Vb can be characterized by influenza virus and has a segmented genome, each segment acting like a template for the synthesis of different mRNA species and encodes a single protein (Lodish et al., 2000). Finally, the class VI viruses are enveloped viruses where the genome contains two identical plus strands of RNA. This type of virus is known as a retrovirus because their RNA directs the formation of DNA molecules which acts as a template for viral mRNA synthesis (Lodish et al., 2000). Retroviruses do not commonly kill their host cell so the cell will continue to replicate and make daughter cells with proviral DNA and bud progeny virions.

Sources

It is difficult to trace viruses through history because they do not leave fossil remains, but molecular techniques are used to compare the DNA and RNA of viruses. There are three different competing theories which try to explain the origin of viruses. The first is the regressive or reduction hypothesis which states that viruses started as independent organisms which became parasites. Over time they shed genes which were useless for their parasitic form and through the years they became dependent on the cells they attack (Crosta, 2017). The second is the progressive or escape hypothesis. This theory states that viruses evolved from sections of DNA or RNA which "escaped" the genes of bigger organisms. This allowed them to become independent and allowed the freedom of moving between cells (Crosta, 2017). The last theory is the virus-first hypothesis, stating that viruses evolved from complex molecules of nucleic acid and proteins around the same time or earlier than the time the first cells appeared on Earth (Crosta, 2017).

Transmission

A virus's main goal is to reproduce so it's offspring can then spread to more hosts and reproduce as well. How well a virus spreads depends on the makeup of the virus (Crosta, 2017). There are a few ways a virus can spread from person to person or from mother to child during pregnancy or delivery, these include: touch, saliva, coughing, sneezing, sexual contact, contaminated food or water, and insects that carry them from one person to another (Crosta, 2017). If a person infected with a virus touches an object some specific types of viruses have the ability to live on the object these objects are known as fomite (Crosta, 2017). The number of hosts is usually a small number.

The reason for this is because the host must have the matching surface receptor which a virus can attach onto (Lodish et al., 2000). Viruses that will infect only bacteria are called bacteriophage and viruses that infect animal or plant cells are called animal viruses or plant viruses (Lodish et al., 2000). There is an exception to the rule and there are some viruses, such as the potato yellow dwarf virus, which can infect both plants and insects which feed on the plant (Lodish et al., 2000). Some types of animal viruses can affect a variety of hosts. The vesicular stomatitis virus is one of them, being able to infect a variety of insects in different types of mammalian cells. But, this is rare and most animal viruses do not cross phyla and some viruses will only infect closely related species such as primates (Lodish et al., 2000).

Lysogenic Cycle versus Lytic Cycle

There are two different ways the virus can replicate, these two ways are the lysogenic and the lytic cycle. The lytic cycle occurs in four events. The first event is absorption where the genetic material is absorbed into the cell, next is penetration where it enters the cell's nucleus, then replication where the genetic code will be turned into mRNA and then will be turned into viral proteins and lastly, release, where these proteins will then be released from the ruptured cell to go and infect other cells (Lodish et al., 2000). The overall outcome is the production of new viral particles and the death of the host cell (Lodish et al., 2000). The lysogenic cycle occurs when the DNA will enter the cell and become integrated into the chromosome but will remain inactive and then is replicated as part of the cell's DNA from one round of replication to the next (Lodish et al., 2000). The integrated DNA from the virus is called the prophage (Lodish et al., 2000). There are certain triggers which can activate the prophage where it will then leave the host cell's chromosomes and enter the lytic cycle (Lodish et al., 2000). There are a few different types of phages and animal viruses which can infect a host cell, start the production of new virion without killing the cell or becoming integrated into the chromosome (Lodish et al., 2000).

Uses

DNA viruses being injected into bacterials and animal cells can help learn about the mechanism of DNA replication since the viruses will depend almost entirely on the cellular proteins (Lodish et al., 2000). There are many different areas in which animal viruses are used, some of these include; cell transformation, AIDS, gene therapy, oncogenes, disease prevention, membrane formation and glycoprotein biosynthesis and intracellular transport (Lodish et al., 2000). Specifically, class I and II are used in studies on DNA replication genome structure, mRNA production and oncogenic cell transformation (Lodish et al., 2000). Class III viruses are useful in research studies on mRNA and translation, class IV and V are useful in studies of membrane formation and intracellular transport and class VI are useful in studies of cell transformation and oncogenes (Lodish et al., 2000).

Conclusion

Viruses are classified as parasites and exist as microorganisms across Earth. They can infect animals, plants, fungi and bacteria, and can cause disease, infection or sometimes have no noticeable reaction. Viruses need a host cell to reproduce and cannot do it on their own. Viruses have two or three parts: genetic material (DNA or RNA), a protein coat called a capsid which will protect the genetic information and sometimes a lipid envelope. There are many different types of viruses such as bacteria specific or animal specific. And animal viruses have six different classes. There are three different competing theories which try to explain the origin of viruses. Additionally, there are many different ways in which a virus can be transmitted to someone and upon transmission they can follow the lytic or the lysogenic cycle. Viruses can be used in many different types of research due to their uniqueness.

WHAT COMMON VIRUSES AND ANTIVIRAL TREATMENTS ARE THERE IN OUR WORLD TODAY?

CAMRYN KABIR-BAHK

In the year 2021, the term virus is all too common. The term can be found anywhere, in online articles, and even in the news. But what exactly are some of the most common viruses in the world? Some may wonder if there are treatments that can also help fight against viruses. This chapter will discuss some of the most common viruses found in the world today and frequently used antiviral therapies that can help fight against common viruses.

Rhinovirus

Remember being a young child and waking up one morning with a stuffy nose? The uncomfortable feeling coupled with chills and a sore throat made it almost impossible to attend school that day. These are some of the tell-tale symptoms of the common cold. Respiratory viruses such as the rhinovirus or coronavirus cause many cases of the common cold.

The human rhinovirus (HRVs) are frequently found in people who have experienced the common cold (Boncristiani et al., 2009). HRVs are non-enveloped, positive-stranded RNA viruses (Boncristiani et al., 2009). This means that the type of genetic information they carry is RNA. The positive-stranded RNA refers to the fact that the RNA located in the virus can be translated to proteins immediately (Lakna, 2018). Whereas with negative-stranded RNA that the virus has is the antisense strand to the viral mRNA (Lakna, 2018). The non-enveloped characteristic of the rhinovirus refers to how the virus does not have a lipid covering. There are two major rhinovirus species, A and B (Boncristiani et al., 2009). However, new-found deletions of genes in the rhinovirus genome may justify a rhinovirus species C soon in the future (Boncristiani et al., 2009). The rhinovirus genome consists of 7.4 kilobases (kb) of single-stranded and uncapped RNA (Boncristiani et al., 2009). The human rhinovirus works by releasing the RNA strand into the cytoplasm of the cell in which the virus has infected (Boncristiani et al., 2009). The RNA is then translated, producing a polyprotein. The polyprotein will then make three precursor proteins; P1, P2, and P3 (Boncristiani et al., 2009). The P1 protein will produce capsid proteins (Boncristiani et al., 2009). The P2 and P3 proteins will help make non-structural proteins (Boncristiani et al., 2009). In addition, a product of the P3 protein is an RNA polymerase that will create a negative-stranded copy of the viral genome which will be used as a template strand to produce positive-stranded RNAs (Boncristiani et al., 2009). The positive-stranded RNA helps create viral proteins or is packaged and sent to new virions (Boncristiani et al., 2009). New HRVs are released from the cell via lysis (Boncristiani et al., 2009). HRVs can survive for many days on surfaces and are resistant to chemicals such as ether, chloroform and ethanol (Boncristiani et al., 2009). However, the human rhinovirus is highly susceptible to halogens (e.g. iodine, chlorine), UV light and pH values lower than five (Boncristiani et al., 2009).

The human rhinovirus is found worldwide and can cause acute respiratory infections in people everywhere. There are recorded traces of the human rhinovirus in populations of people located in Alaska, the Kalahari Desert and even isolated tribes in the Amazon (Boncristiani et al., 2009). The human rhinovirus is also incredibly prevalent in North America and Western Europe (Boncristiani et al., 2009). In fact, HRVs account for approximately 80% of cases of fall common colds found in adults in the United States (Boncristiani et al., 2009). The human rhinovirus transmits at its highest rates when there are reasonably humid conditions (Boncristiani et al., 2009). Crowded conditions can also increase the spread of HRVs. Transmission of the rhinovirus will usually occur when two people are in very close contact and engage in hand-to-hand contact, although it can still spread by droplets in the air (Boncristiani et al., 2009). Usually, the rhinovirus will almost certainly infect an individual if the virus makes its way to the nasal cavity (Boncristiani et al., 2009).

The human rhinovirus will incite symptoms of the common cold by replicating within ciliated cells located in the nose (Boncristiani et al., 2009). Once the cells are infected, the expression of inflammatory mediators, cytokines and chemokines will begin (Boncristiani et al., 2009). Cytokines and chemokines are proteins that control the immune response within the body (Ramesh et al., 2013). In addition to the expression of these proteins and inflammatory mediators, the parasympathetic nervous system will become activated and lead to an individual experiencing common cold symptoms (Boncristiani et al., 2009). The activation of cytokines and chemokines will help induce common cold symptoms.

Frequent symptoms of the common cold are a "stuffy" nose, nasal discharge, coughing, sneezing, headaches, chills and a sore or irritated throat (Boncristiani et al., 2009). Less frequent symptoms will include pressure in the ears and a fever. Typically, symptoms will last for about a week (Boncristiani et al., 2009). In approximately a quarter of common cold cases, symptoms can last for about 14 days (Boncristiani et al., 2009). Toddlers will sometimes appear to be asymptomatic (Boncristiani et al., 2009). Although there are more than one human rhinovirus species, there is no correlation between certain symptoms and a specific virus species (Boncristiani et al., 2009).

Influenza Virus

If asked, almost everyone will say that they have experienced the flu once in their lifetime. The influenza virus is amongst one of the most common viruses worldwide. Each year, the influenza virus infects 10% of adults and 20% of children worldwide (Peteranderl et al., 2016). Furthermore, this common virus is responsible for approximately 250,000 to 500,000 deaths each year (Nguyen, 2020). Each year, the influenza virus has a significant impact on healthcare worldwide.

There are two strains of the influenza virus that are prevalent and cause common infections in humans; influenza type A and B (Arbeitskreis Blut, 2009). However, influenza types C and D are strains that do exist (CDC, 2019). Influenza type C will typically induce very mild symptoms. Influenza type D is not prevalent in humans; it usually only infects cattle and other animals (CDC, 2019). Influenza viruses are negative-sense stranded RNA viruses with an enveloped lipid covering (Arbeitskreis Blut, 2009). Influenza A and B genomes consist of eight single-stranded RNA pieces (Bouvier & Palese, 2008). The virus infects epithelial cells present in the respiratory tract. They infect by recognizing N-acetylneuraminic (sialic) acid present on the surface of an epithelial cell (Bouvier & Palese, 2008). It will then attach its HA spike proteins to the sialic acid (Bouvier & Palese, 2008). The virus is then endocytosed and brought into the host cell (Bouvier & Palese, 2008). The low pH in the cell is crucial to the mechanism of infection which influenza viruses use. Low pH will cause the HA proteins to change in shape and expose a peptide which will help merge the viral envelope to the cell's internal membrane (Bouvier & Palese, 2008). This process will help open the virus and release viral ribonucleoproteins into the cell's cytoplasm (Bouvier & Palese, 2008). The ribonucleoprotein gets brought to the cell's nucleus, where the viral RNA gets synthesized (Bouvier & Palese, 2008). The negative-sense stranded viral RNA is used as a template to synthesize positive-stranded RNA (Bouvier & Palese, 2008). The positive-stranded RNA acts as mRNA templates used to synthesize viral proteins (Bouvier & Palese, 2008). Once the mRNA synthesis is completed, it will be exported and translated in the host cell's cytoplasm (Bouvier & Palese, 2008). Proteins such as the HA spike proteins are synthesized from the mRNA and shipped out of the nucleus. After protein synthesis, they are transported to the Golgi apparatus for post-translational modification and packaging (Bouvier & Palese, 2008). The ability for an influenza virus to effectively infect a cell is maintained as long as the virus has a complete viral genome (Bouvier & Palese, 2008). Viral proteins and RNA are assembled and packaged into a new viral particle and then buds off the host cell's membrane (Bouvier & Palese, 2008).

Influenza A and B are responsible for flu epidemics that occur each year (CDC, 2019). Epidemics usually occur during December to April in the Northern Hemisphere and between June and September in the Southern Hemisphere (Peteranderl et al., 2016). The conditions such as low humidity during the winter months work to the virus's advantage, prolonging its transmission (Peteranderl et al., 2016). The spread of the virus usually occurs through droplets (Peteranderl et al., 2016). Seasonal influenza A will usually induce a mild upper respiratory tract illness or will appear asymptomatic in the individual (Peteranderl et al., 2016). Symptoms such as coughing, headaches, chills and fever are usually observed within one or two days of incubation (Peteranderl et al., 2016). Symptoms might persist for up to 8 days. Rarely, complicated diseases and conditions such as pneumonia might be prompted (Peteranderl et al., 2016).

Every 30 years, a seasonal epidemic will turn into a pandemic (Peteranderl et al., 2016). Pandemic strains of influenza A usually have some sort of genomic change causing more people to be susceptible (Peteranderl et al., 2016). New strains will also lead to a higher mortality rate and more serious symptoms such as vomiting or diarrhea (Peteranderl et al., 2016).

Rotavirus

The rotavirus is the leading cause of viral gastroenteritis, or an illness more commonly known as the "stomach flu" (Crawford et al., 2017). They are double-stranded RNA, non-enveloped viruses (Crawford et al., 2017). The rotavirus genome consists of 11 segments of RNA, which code for twelve viral proteins, half of which are structural and the other half are non-structural proteins (Crawford et al., 2017). Rotaviruses infect mature and non-dividing cells in the intestinal lining (enterocytes) located in the small intestine (Crawford et al., 2017). The virus attaches to receptors located on the surface of a host cell, and viral proteins will interact with co-receptors, thus mediating the entry of the virus (Crawford et al., 2017). Once the virus enters the host cell, low calcium levels will cause the removal of the virus's outer capsid layer (Crawford et al., 2017). Transcriptionally active double-layered particles are released into the cell's cytoplasm (Crawford et al., 2017). Viral mRNA is also released into the cytoplasm and used as a template to replicate the viral genome (Crawford et al., 2017). Newly replicated genomes are packaged into double-layered particles inside of viroplasms (Crawford et al., 2017). Viroplasms are structures made of cellular and viral proteins that need lipid droplets to be fully formed (Crawford et al., 2017). In the endoplasmic reticulum, outer capsid proteins get added to enveloped virus particles; however, the envelope is lost shortly after (Crawford et al., 2017). Mature virus particles are usually released from the host cell via lysis (Crawford et al., 2017).

The most common symptom of the rotavirus is diarrhea, and it usually lasts between 3 to 8 days. Other symptoms may include abdominal pains, fever and vomiting. Symptoms will usually appear about two days after the initial infection.

The rotavirus is incredibly common and dangerous, especially amongst young children. In 2003 alone, about 114 million cases of rotavirus were reported worldwide amongst children under the age of 5 years old, and almost 2.3 million of those cases required hospitalization (Crawford et al., 2017). Although children everywhere are affected by rotavirus, more than 90% of fatal cases come from children living in low-income countries (Crawford et al., 2017). The virus is frequently transmitted by close contact with an infected person. Contaminated objects play a prominent role in spreading the virus in settings such as hospitals, schools or day-cares (Crawford et al., 2017).

Human Immunodeficiency Virus

The human immunodeficiency virus (HIV) has had a massive impact on society and the healthcare industry. HIV is a virus that attacks the immune system and can lead to acquired immunodeficiency syndrome (AIDS) if it is left untreated. HIV is a type of retrovirus with two significant subtypes; HIV-1 and HIV-2 (Waymack & Sundareshan, 2020). HIV-1 is the most common subtype of HIV and is responsible for causing AIDS (Waymack & Sundareshan, 2020).

The human immunodeficiency virus is spherical with a lipid layer (Arbeitskreis Blut, 2016). HIV attacks the immune system, and in doing so, it infects white blood cells. HIV attaches to the host cell using viral glycoproteins (Waymack & Sundareshan, 2020). The viral genome consists of two uniform pieces of single-stranded RNA located in the virus's core (Arbeitskreis Blut, 2016). The genome is generated using an enzyme called reverse transcriptase. Reverse transcriptase works by reverse transcribing the single-stranded RNA into double-stranded DNA. The virus then takes over the cell's machinery and incorporates its viral DNA into the host cell's DNA, producing more viral proteins and genetic material (Waymack & Sundareshan, 2020). As more copies of HIV get replicated, they leave the host white blood cell and eventually, the cell will die (Waymack & Sundareshan, 2020). New HIV viruses will infect more white blood cells, thus producing more human immunodeficiency viruses.

HIV/AIDS is considered to be a global pandemic and crisis. The virus was identified in the early 1980s, and since then, approximately 39 million people have died due to infection (Waymack & Sundareshan, 2020). Currently, more than 35 million people are living with the disease (Waymack & Sundareshan, 2020). Although the virus has become more prevalent in society now, the number of yearly cases reported has begun to decline since the 1990s (Waymack & Sundareshan, 2020).

HIV is typically transmitted through sexual intercourse, breastfeeding or childbirth, sharing intravenous needles or through blood transfusions (Waymack & Sundareshan, 2020). HIV transmitted sexually can be detected two days after the initial infection in lymphatic tissue (Arbeitskreis Blut, 2016). HIV can be detected in the lymph nodes about six days after infection, and within two weeks, it can be detected throughout the entire body (Arbeitskreis Blut, 2016). When HIV is transmitted through blood, it can be detected within 5 or 6 days (Arbeitskreis Blut, 2016). HIV transmission between mother and child can occur as early as the 12th-week of gestation (Arbeitskreis Blut, 2016). However, transmission in over 90% of cases that arise in newborn babies occurs in the final trimester (Arbeitskreis Blut, 2016). About 3-6 weeks after infection, typical symptoms such as lymph node enlargement, fever, rashes with lesions, gastrointestinal symptoms and fatigue are observed (Arbeitskreis Blut, 2016). This initial phase can last for 2-6 weeks and is followed by a mostly asymptomatic phase that can last for years (Arbeit-skreis Blut, 2016). If treated properly, it is possible for people infected with HIV to live a relatively normal life. However, if left untreated, HIV will continue to replicate and kill white blood cells, rendering the immune system helpless. Individuals who do not seek treatment will develop AIDS within ten years and will likely die after two years of having AIDS (Waymack & Sundareshan, 2020).

Common Influenza Antivirals

Most viruses, except human immunodeficiency virus, influenza, herpes, and hepatitis, do not have an antiviral treatment (Razonable, 2011). Amongst these antiviral treatments for HIV and influenza are some of the most common antiviral therapies.

There are two main antiviral drugs that are commonly used and approved by the FDA to combat seasonal influenza viruses; oseltamivir and zanamivir (Razonable, 2011). Oseltamivir works by inhibiting neuraminidase enzymes from cleaving sialic acid (Razonable, 2011). Sialic acid is crucial to the function and replication of influenza viruses, as it will block the release and production of newly synthesized viruses (Razonable, 2011). Oseltamivir is taken orally, usually in the form of a pill. After one hour, the drug reaches its highest concentration within the body (Razonable, 2011). The use of oseltamivir should ideally begin within 48 hours of observing symptoms, and continued usage of the drug should last up to 5 days (Razonable, 2011). After use, the drug gets metabolized and excreted (Razonable, 2011). Oseltamivir is used to treat adults and children older than one year who are suffering from the flu (Razonable, 2011). Oseltamivir can also be used to fight against pandemic strains of influenza (Razonable, 2011). In this case, the drug should be taken for a minimum of 10 days and a maximum of 6 weeks (Razonable, 2011). Individuals who take oseltamivir can expect to experience some side effects. This drug's common side effects include vomiting, nausea, insomnia, abdominal pains, and insomnia (Razonable, 2011). Some cases of neuropsychiatric side effects such as hallucinations and delirium have been reported (Razonable, 2011). Oseltamivir is not always fully effective. There have been reports of cases where influenza A appeared to be resistant to the drug (Razonable, 2011). Resistance to oseltamivir is likely caused by mutations in the genes which code for neuraminidases (Razonable, 2011).

Another prevalent antiviral treatment for influenza viruses is the drug zanamivir. This drug acts very similarly to oseltamivir. Zanamivir also works by inhibiting the function of neuraminidase (Razonable, 2011). However, zanamivir is not effectively absorbed when taken orally, so it is inhaled (Razonable, 2011). High concentrations of zanamivir occur 1 to 2 hours after inhaling and are present in the respiratory tract (Razonable, 2011). Absorbed zanamivir does not get metabolized; it is released from the body unchanged through urination (Razonable, 2011). Unabsorbed zanamivir is removed from the body through fecal excretion (Razonable, 2011). Zanamivir is given to an individual two days after symptoms begin and usage continues for five days, similar to oseltamivir (Razonable, 2011). Individuals take two doses of zanamivir each day (Razonable, 2011). Not many side effects occur after using this drug, except for a mild headache or gastrointestinal symptoms (Razonable, 2011). Neuropsychiatric effects similar to oseltamivir rarely occur (Razonable, 2011).

Common HIV Antiviral Treatments
HIV is a difficult virus to treat. Unlike some of the other viruses discussed in this chapter, individuals who acquire this virus will have it for the rest of their life. Fortunately, the medical field has been researching antiviral treatments that can help combat HIV. Nowadays, antiretroviral therapy is used to fight against HIV. Multiple antiretroviral drugs are used in a "cocktail,"

which is a mixture of drugs. Antiretroviral drugs target HIV and attempt to inhibit its ability to replicate, ensuring more white blood cells are not infected. The CDC has approved six classes of antiretroviral drugs, which are separated based on the phase of HIV replication it targets (Kemnic & Gulick, 2020). Standard drug cocktails given to patients include two nucleoside reverse transcriptase inhibitors, either a protease or integrase inhibitor and a non-nucleoside reverse transcriptase inhibitor (Kemnic & Gulick, 2020). HIV can easily acquire resistance to drug treatments, thus using multiple drugs at once stalls resistance the virus may gain. Over 20 FDA approved HIV medications and drug combinations depend on the individual (Kemnic & Gulick, 2020).

Nucleoside reverse transcriptase inhibitors work by competing against natural nucleotides for integration into growing viral DNA chains (Kemnic & Gulick, 2020). However, nucleoside reverse transcriptase inhibitors do not have a 3' hydroxyl; therefore, once it is incorporated into the DNA chain, other nucleotides will not be able to attach (Kemnic & Gulick, 2020). This will immediately terminate the growth of the viral DNA chains (Kemnic & Gulick, 2020). Some side effects that come with the use of nucleoside reverse transcriptase inhibitors include anemia, rashes, fever, lactic acid build-up, diarrhea, sore throat, jaundice and abdominal pains (Kemnic & Gulick, 2020). Non-nucleoside reverse transcriptase inhibitors are another class of drugs that are very common in HIV treatment. These drugs work by directly binding to and inhibiting the reverse transcriptase enzyme (Kemnic & Gulick, 2020). This process will ensure that viral RNA does not get turned into DNA, effectively terminating the HIV replication cycle (Kemnic & Gulick, 2020). Some side effects of the drug include rashes, nausea, vomiting, depression, blisters and mouth sores (Kemnic & Gulick, 2020). Protease inhibitors block the cleavage of precursor proteins used to produce HIV proteins (Kemnic & Gulick, 2020). Some of the side effects of protease inhibitors are kidney stones, abdominal pains, swelling, rashes, lightheadedness, heartburn and pancreatitis (Kemnic & Gulick, 2020). Integrase inhibitors are another typical class of drugs. Protease inhibitors or integrase inhibitors are commonly used in drug cocktails (Kemnic & Gulick, 2020). Integrase inhibitors prevent the viral genome from being inserted and incorporated with the host cell's DNA (Kemnic & Gulick, 2020). Blocking this process will stop HIV from using the host cell's machinery to create more HIV. Common side effects of integrase inhibitors are skin blisters, fatigue, jaundice, abnormal dreams, flatulence and loss of appetite (Kemnic & Gulick, 2020). Patients can take most HIV medications orally and in the form of a pill (Kemnic & Gulick, 2020). Patients participating in an antiretroviral treatment must avoid using illicit drugs (Kemnic & Gulick, 2020). For many patients, antiretroviral treatment is a lifelong commitment.

Conclusion

Practically all humans have been affected by a virus at some point in their life. Whether it is something mild like the rhinovirus or something more severe such as the human immunodeficiency virus, viruses are everywhere. Viruses and their impact on the healthcare system have always dominated global news, especially in 2021. Whether it is ICU beds filling up or long hours of research going into developing new antiviral treatments, viruses can sometimes be catastrophic. Although there are hundreds of viruses, it is imperative to be educated on common viruses and antiviral treatments used. Viruses are spread but so is information, so be a responsible citizen and contribute to the latter, not the former.

WHAT SCIENCE IS INVOLVED IN STUDYING VIRUSES?

CHRISTINA NGUYEN

Classification and Nomenclature

In the previous chapters, we note that all viruses contain a single type of nucleic acid: either RNA or DNA. Thus we categorize them according to the type of nucleic acid they possess: "RNA virus" or "DNA virus." A circular genome is found in some viruses, while a linear genome is found in others. A virus's genome can be single-stranded or double-stranded, and the type of genome it has determines how it replicates. Viruses are classified as helical, icosahedral, or complex in shape. When viewed under an electron microscope, helical viruses appear cylindrical and can be either short and rigid or long and filamentous. Under an electron microscope, icosahedral viruses appear to be circular, but their surface is more like that of a soccer ball, with 20 flat triangles arranged around the perimeter. Complex viruses are more intricately designed and have a wider range of shapes. Viruses also have unique protein components that enable them to bind to receptors on host cell surfaces.

Unfortunately, the process of viral classification and nomenclature is far from uniform worldwide. Various methods and styles are still in popular usage. Fortunately, the International Committee on Taxonomy of Viruses (henceforth referred to as the ICTV,) is an entire international committee devoted to classifying viruses. A virus species is classified as a population of viruses that share a pool of genes that is distinct from the gene pools of other viruses, according to this scheme. Viruses can be classified into organisms even though we do not consider them to be alive. A genus is a classification system for viruses.

A subfamily is a set of genera (plural for genus) that are related. A family is a set of closely related subfamilies. Finally, an order is a set of associated families. The majority of virion and genome traits are used to classify them into families. The following are the criteria: what is the virion's size and shape? Is there an envelope with the virion? Electron microscopy is commonly used to assess all of this. Following that, there are all of the structural proteins. What is the virus's molecular weight?

Finally, there are genome properties to consider. Is it a nucleic acid of some sort? Is it a single-stranded or a double-stranded cable? Is it positive-sense or negative-sense if it is single-stranded? Note that a "positive-sense" viral RNA can be immediately translated by the victimized cell; on the other hand, "negative-sense" viral RNA takes a bit more effort to copy over – it must be converted over the "positive-sense" before it can be translated (Lumen Candela, n.d.). Another question is, what is the nucleotide sequence? These are the parameters that are taken into account. Virus names, with the exception of names of virus families ending in the italicised suffix -viridae, have no clear sequence. Others are named after the virus's emergence, others after the geographic area in which they were found, still others after the disease they cause, and still others after their mode of transmission. While bacteria are usually identified by their genus and species names, viruses are identified solely by their species name. To make matters worse, many virologists use colloquial terms for viruses. For example, the current pandemic (at the time of writing) has been caused by the COVID-19 virus (informal name), or in long form, the coronavirus disease 2019 (having originally emerged in 2019). But that tells us little about the virus itself. We have to look at the more formal name of SARS-CoV-2, which at least tells us that it is a 'severe acute respiratory' illness, i.e. that it affects our breathing systems (e.g. lungs).

Observations under microscopes

1.   The electron microscope

Some viruses are visible under an electron microscope, to detect pathogens. The specimen is put onto electron microscope grids, and then placed under the microscope to see the structure of both the virus and the infected cells. The qualities of these two can often help identify which pathogen is present. This method is particularly useful when doctors have a patient exhibiting symptoms but have no diagnosis of which disease it is yet. The electron microscope is very useful in giving an overall look at the virus's makeup and nature.

2.   The light microscope

Light microscopes are different from electron microscopes; they fundamentally view different wavelengths from the electromagnetic spectrum. Light microscopes are able to see larger items, while electron microscopes are able to see down to the thousandth of a millimeter. Light microscopes unfortunately cannot observe viruses directly, because viruses are much smaller than the wavelength that the light microscope allows.

Take the coronavirus, for example. A light microscope cannot see it because viruses are way too small and below the resolution limit of what light can resolve; but this is a tricky situation. Theoretically speaking, if the contrast was high enough, it is possible to see even particles that are smaller than the limit of the microscope's resolution. Here I will outline the method used to do this.

The microscope is a diffraction-limited device, and this means that if you go up with the magnification too much, you will see that the image under the lends becomes more and more blurry, and we cannot see any details of our specimens on the slide or Petri dish at all! However, the blurry area around each specimen can help us actually detect the virus. Let us set an example: two objects are under a light microscope, and when we look at them, we see two blurry areas, that is, if the objects are far from each other and separated well (Davidson, n.d.). These are "contrasted" with each other enough that thee "airy circles" containing miniscule viruses are still visible under a light microscope, despite them being smaller than the wavelength permitted. What if the two circles were close together? Then it is not highly contrasted and we cannot see the viruses' locations clearly. Another case is if the viruses are at the same distance as the first case mentioned here, but the resolution is much lower. Then the two airy circles would be one big fuzzy lump - and we cannot see very clearly at all! An electron microscope would be needed to observe anything at all of the coronavirus. In other words, the resolution is defined as the minimum distance that you can have and the objects are still visible as two separate objects under any type of microscope (and light and electron microscopes are certainly not the only types out there - for example, astronomical telescopes use the same principles too). Of course, you are going to be able to see objects that are smaller than the "resolution limit" if, and only if, the contrast is sufficient. Naturally you can also expect to see no details on the objects, only the separation (Abramowitz & Davidson, n.d.).

Virus surveillance and security

The motivation behind funding virology studies is to protect humans from dangerous outbreaks. In this section, we will be discussing specifically the current case of the coronavirus, which necessarily includes discussions about politics and nationalism. During the early stages of the pandemic, most Western countries failed to keep up with the demand for COVID-19 testing. It was difficult to determine how far the virus had spread or how it would spread in the future without research, which meant we were going into an unknown future with no plans.

Now, North America is dealing with a similar problem involving a particular form of testing: genetic sequencing. Genetic sequencing decodes the genome of virus in patient samples, unlike the COVID-19 test, which diagnoses infection. Knowing the virus's genome sequence aids researchers in deciphering two main points: how the virus mutates into variants and how it spreads from person to person (Oschner Health, 2020).

Prior to the COVID-19 pandemic, genomic surveillance was primarily used to perform small studies on antibiotic-resistant bacteria, investigate outbreaks, and monitor influenza strains. We conduct these types of experiments every day in our laboratories as genomic epidemiologists and infectious disease researchers, trying to figure out how the coronavirus is developing and spreading across the population. Genomic surveillance plays a critical role in bringing the pandemic under control, particularly now that new coronavirus variants of concern are emerging, such as the British (B.1.1.7) and Brazilian (P.1) variants that we now see (CDC.gov, 2021).

Several methods are used to track the virus's spread.Genome sequencing entails determining the sequence of nucleotide molecules that make up a virus's genetic code. A series of about

30,000 nucleotides make up the coronavirus genome. Errors are produced every time the virus replicates. Mutations are the result of errors in the genetic code.The majority of mutations have little effect on the virus's work. Others may be relevant, particularly if they encode essential elements like the coronavirus spike protein, which serves as a key for the virus to enter human cells and trigger infection. Spike mutations can affect the virus's infectiousness, the severity of infection, and how well current vaccines protect against it. Researchers are looking for mutations that differentiate virus specimens from one another or fit recognized variants.

Scientists can use the genetic sequences to track how the virus spreads in the population and in hospitals. If two individuals have viral sequences that are identical or have only minor variations, it is likely that the virus was passed from one to the other or came from a common source. The genetic sequences can be used by scientists to track how the virus spreads in the population and in health-care facilities. If two individuals have viral sequences that are identical or have very few variations, it is likely that the virus was transmitted from one to the other or from a common source. On the other hand, if the sequences vary significantly, these two people did not contract the virus from each other. This type of data allows public health officials to adapt programs and guidelines to the needs of the general public.

In health-care settings, genomic surveillance is also relevant. For example, our hospital employs genomic surveillance to identify outbreaks that would otherwise go undetected by conventional methods. Surveillance will give you a heads-up. But how do scientists know if new variants are appearing and whether or not people should be concerned? Consider the B.1.1.7 variant, which was first discovered in the United Kingdom and is under strict genomic surveillance. In the United

Kingdom, researchers discovered that a specific sequence of many modifications, including the spike protein, was on the increase. And with the government shut down, this strain of the virus was spreading faster than its predecessors.

Scientists dug deeper into the genome of this variant to see how it was able to circumvent distancing guidelines and other public health interventions. They discovered 69-70 and N501Y mutations in the spike protein, which made it easier for the virus to infect human cells. According to preliminary studies, these mutations resulted in a higher rate of transmission, implying that they spread much faster than previous strains.

Vaccine creators and other scientists have used this genetic data to see how the new variations affect the vaccines' effectiveness. Preliminary testing, which has not yet been peer-reviewed, has shown that the B.1.1.7 strain is still vulnerable to current vaccines. Other versions, such as P.1 and B.1.351, which were first discovered in Brazil and South Africa, respectively, can evade some of the vaccine's antibodies (NCBI, n.d.).

How is a genomic surveillance system set up? A comprehensive genomic surveillance program is needed to detect variants of concern and establish a public health response to them.

That means scientists would sequence virus samples from around 5% of the total number of COVID-19 patients, chosen to represent the populations most vulnerable to the disease. Without this genomic data, new variants could spread rapidly and undetected across the country and around the world. So, how does the United States fare in terms of genomic surveillance? In terms of the number of SARS-CoV-2 genomes sequenced per number of cases, the United States ranks 34th, well behind other developed countries. The Centers for Disease Control and Prevention (CDC) in collaboration with other federal agencies, is increasing genomic surveillance capability in the United States by collaborating with private laboratories, state and local public health laboratories, academia, and others (The Conversation, 2021). Laboratories must procure samples from a variety of places, including public health laboratories, hospitals, clinics, and private research facilities. Bioinformaticians use sophisticated programs to recognize essential mutations after the sequencing test is completed. After that, public health experts combine genetic and epidemiological data to figure out how the virus is circulating. All of this necessitates an effort in teaching people to work together to complete these tasks.

Finally, an effective genomic surveillance program must be fast, allowing data to be made publicly accessible right away to help public health officials and vaccine manufacturers make real-time decisions. One of the public health tools that will help get the current pandemic under control and prepare the United States to respond to possible pandemics is a program like this.

The year 2021 appears to be the year of COVID-19 models. Scientists have discovered many fast-spreading variants in the last two months, prompting government restrictions in a number of countries — and new variants are being discovered more regularly (CDC.gov, 2021).

The pandemic has ushered in a new age of genomic surveillance, in which scientists are monitoring viral genomic changes at a never-before-seen pace and size. However, global surveillance is patchy, particularly in the United States, which has the world's largest COVID-19 outbreak, and several low- and middle-income countries. Concerning variants are likely spreading undetected in these areas, according to scientists.

Looking at surveillance networks, the ability to track mutations and variants of concern as they emerge is dependent on the sequencing and sharing of enough genomes. Over 360,000 SARS-CoV-2 genomes were sequenced and deposited on GISAID, a non-profit online database for sharing viral genomes, in the last year. The sequences on GISAID have a broad geographic distribution, spanning more than 140 countries. However, the majority of countries have only uploaded a limited number of sequences (NCBI, n.d.). Surveillance networks vary in size from massive national programs to small community-based operations.

Conclusion

There are many scientific methods involved in studying viruses, and certainly, these involve various fields of science as well. This includes biology and social modeling – and the field is rapidly expanding to accommodate future needs to protect the safety of humans everywhere.

WHAT CONTROVERSY IS THERE SURROUNDING VIRUSES?

BRIANNA BEDRAN

In the late 1980s hopes were dashed that emerging infectious diseases were a problem of the past. The Emerging Viruses: The Evolution of Viruses and Infectious Diseases conference held in Washington DC 1989 raised concerns of emerging viruses and viral illnesses due to evolutionary responses to human-made developments in viruses environments (Washer, 2014) Researchers, professors, and key-note speakers at the medical conference used examples of the human immunodeficiency virus (HIV) and the exotic viral illness ebola to justify how viruses are constantly emerging. Given the unprecedented population growth of the modern world, concerns of the ability of these viruses to manifest in these susceptible large demographics was one of the major concerns (Washer, 2014).

Lack of basic sanitation in the developing world contributed to cases of cholera and dengue fever, making it a major concern for viruses' ability to manifest in these environments. In both the developed and developing world, the large under-vaccinated population is a contributor to virus development. Many concerns of the modern world were brought up, such as industrialization and thus deforestation, and international travel as a contributing factor. The most crucial problem, however, is the dismantling and breakdown in public health measures (Washer, 2014). Political parties played a role in this, as republican administrations maintained pressure to constrain budget requests. The total funding into research for emerging infectious diseases was $1.7 billion in 2005, it is not clear whether this was in direct response to the concerns raised in the conference, but overall EID was successfully established as an official branch of science (Washer, 2014).

Despite the significant financial investments in research and crucial concerns brought up in the conference, there are multiple controversies and misconceptions that exist within the discourse of viruses. Using examples of HIV, Zika virus, and applying patterns with the most current case of COVID-19, this chapter will explore these controversies. Virus controversy demonstrates concerns of lack of preparedness and differing policy responses among governments, immunization and virus conspiracies, and stigmatizations of marginalized groups.

Government Preparedness and Policy Response

Roughly a year prior to the COVID-19 pandemic, Inglesby and Toner reported that western health systems and their abilities to control and deter emerging infectious diseases have not been given merely enough attention. Similar to the discourse brought up by Washer, both authors mention that there is relatively limited public investments to strengthen and advance epidemiological surveillance systems. Additionally, Ingesby reports that there are still not licensed vaccines for many deadly viral pathogens that have emerged naturally over the past forty years. The U.S vulnerability to such diseases became even more evident when hospitals grappled with the limited amount of Ebola patients in 2014. Using an exercise simulating the spread of a severe new pandemic involving a virus called Clade X, Inglesby, Toner, and other colleagues at the Johns Hopkins Center for Health Security exposed the realities of the United States lack of preparedness for a real deadly pathogen (Inglesby, et al., 2018). This exercise made it apparent that the only way to stave off a catastrophic outcome would have been a global public-health system capable of rapidly detecting a nascent outbreak and responding vigorously before it could become a pandemic. As evident in the Clade X simulation, these researchers concluded that such preparedness doesn't exist today in the United States (Inglesby, et al., 2018).

Their estimations have been demonstrated in the case of COVID-19, where the United States approach and management of COVID-19 involved a delayed response that focussed mainly on patient-centered strategies and case management in the hospital settings. Their mistakes in dealing with COVID-19 is also explained by them underestimating the negative impact of the viral spread in the community, while ignoring the advice of their scientific advisors. By the time they came to implement total lockdown and a substantial number of tests, their response was very late in the game (Shokoohi, et al., 2020). Drew Altman reports that the United States has by far the highest case and death count globally. The historic neglect and underfunding of the state and local public health system have contributed to the weak US response. Lack of preparation played a role in their mistakes, however Altman notes the underestimation of the dangers of COVID-19 preventing proper responses were taken when the Trump administration was in office (Altman, 2021). Either way, despite the Emerging Viruses conference being held in the United States, which should have made them aware of the dangers, their management of COVID-19 displays a lack of pre-planned action for epidemiological control.

Lack of preparedness can be identified in South Asia as well, however their response was much more swift than in the case of the United States. South Asian nations do not have guidelines or national plans that can manage the surveillance and control of multiple zoonotic pathogens of concern for the public's health. Countries in this region also have a shortage of health workers, and less than one field epidemiologist per 200,000 people (Babu, et al., 2020). In addition, there is a high burden of non-communicable diseases in South Asia that multiplies their difficulties with combating present and future health crises. Their advantage of having a larger younger population demographic in South Asia may not be enough for the rising burden of non-communicable diseases and lack of priority setting for improving health systems. Thus, as measured by the Global Health Security Index, South Asian countries generally had inadequate pandemic preparedness (Babu, et al., 2020). However, their early lockdown and timely intervention for prevention and response, measures the United States did not take, demonstrated much more success in South Asia compared to other countries in containing the spread of the virus. Despite the major investments in research and raising awareness of emerging diseases, COVID-19 demonstrated confusion of proper responses and planning among nations, establishing a notable virus controversy. The variation in different countries' responses, United States slow response underestimating dangers, versus South Asia's swift response, has been an ongoing topic among many more scholars, and news outlets in the past year than we have referenced. It has raised concerns and doubts in our knowledge of viruses, and the plans to counter the spread. Especially with the case of COVID-19, the controversy raised here in lack of preparedness and differing responses will remain a major controversy in the study of viruses.

Virus Conspiracies

More controversies occurring prior to COVID-19 is the rise of virus and immunization conspiracy theories, manifesting most prominent in the digitalized age on social media platforms. This is exemplified by the Zika virus outbreak of 2015-2016 which sparked many controversies on social media regarding its authenticity. When it was linked to a rapid increase in microcephaly cases in Latin America, the mosquito-borne Zika virus was the topic of many international media outlets starting in early 2016. The expert consensus is that the virus originated in Africa, and is carried by the Aedes aegypti mosquito, and can potentially lead to microcephaly and other fetal abnormalities if a woman is infected early in pregnancy (Wood, 2017). A study that investigated the characteristics on discourse of conspiracy theories about the virus outbreak during its prime presence in the media analyzed over twenty-five thousand tweets discussing the virus, finding the words "hoax" and "false flag" appearing often. The type of conspiracy theories found in these tweets take on multiple forms. One theory is that the virus is a bioweapon rather than a natural occurrence. Another is that the virus is actually harmless and the microcephaly epidemic is instead caused by pesticides, genetically modified mosquitoes, or vaccine side effects. Another claim is that Zika vaccine development efforts are part of a broader plan for global depopulation (Wood, 2017). Many of these claims also have clear links to other conspiracy theories, for example the biotechnology corporation Monsanto or the Rockefeller business empire are sometimes presented as the most likely perpetrators of the conspiracy is a common theme in conspiracy discourse. The posts regarding Zika virus conspiracy theories demonstrate that relative to debunking messages, the theories spread through a more decentralized network, are more likely to invoke supposedly knowledgeable authorities in making arguments, and ask more rhetorical questions (Wood, 2017). For instance, conspiracy theories claiming discreet government persecution of African-Americans, similar to the idea that HIV/AIDS is a bioweapon created for racial genocide were popular in African-American communities. This is particularly among people who felt victimized and alienated from institutions like government and politics (Wood, 2017). It is notable, however, that these feelings among African-American's are not irrational in nature like some of the other theories mentioned, as the government does have a history of neglecting this community. For the most part, conspiracy theories about a particular event arise from people's desire for sense-making, and that this need is exaggerated when the event in question is both serious and self-relevant. Wood suggests that this is essentially the same pattern under which rumors tend to arise (Wood, 2017).

The COVID-19 virus invoked controversies that range from political goals and the blame-game for the pandemic. A recent study measured variables possibly associated with little feelings of worry about COVID-19, and notions that China is responsible for the virus outbreak. This study observed authoritarianism, conspiracy beliefs, gender, and consistency of handedness as predictors of nine likert scale questions gauging attitudes, behavior, and beliefs regarding the virus. To reduce the number of dimensions and tests that can yield a type one error, the nine items were then submitted to a principal components analysis which yielded a "concern about COVID" factor and a "Blame for China" factor (Prichard et al., 2020). Results revealed that conspiracy beliefs are often coupled with a stronger belief in China's responsibility for the pandemic. Additionally, women expressed greater degrees of concern about their own and others' health and about the financial wellbeing of others, meanwhile men demonstrated less concern about the virus overall. Results also proved that authoritarianism is generally associated with less concern about the virus (Prichard et al., 2020).

In addition, social media bots are actively being used to spread misinformation about the COVID-19 virus. Journalism findings reveal that the collection of social media data can be effectively used by political organizations to target propaganda to audiences from key demographics based on behavioural and personality data collected online (Prichard et al., 2020). Public transparency about the type of personality variables that can be targeted may give individual citizens and public policymakers some means of defending themselves against this kind of manipulation. As such, psychological scientists could play an important role in understanding which personality factors contribute to negative health behaviors during the outbreak (Prichard et al., 2020). Prichard and Christman suggest that if policymakers want to design effective messages that encourage individuals to take virus science serious and take the proper precautions to prevent illness, they must have an idea of which personality types are more susceptible to ignoring the recommendations of experts and believing in the conspiracies mentioned above (Prichard et al., 2020). The Pew Research Center also conducted a similar study on COVID-19 controversies that report the effectiveness of increasing polarization of attitudes toward the pandemic. The findings conclude that 38% of respondents believe the seriousness of COVID-19 is being exaggerated, and 36% of all respondents across the study reported believing that it is definitely or probably true that the outbreak was a planned conspiracy (Prichard et al., 2020).

The COVID-19 virus and thus pandemic requires a need for people to make sense out of such a crazy situation, making it difficult to rationalize that such a tragedy can simply be caused by mother nature. This desire for reasonableness leads to these controversial conspiracies, spreading misinformation and a lack of concern for such a die circumstance. We can also see similar patterns of conspiracy theories between COVID-19 and the Zika virus outbreak arising and manifesting especially in the online world.

Misconceptions and Stigmas of Viruses

Peter Washer describes how notions of high risk groups for viruses are often used to mask stereotypes. This controversy can be identified in the case of HIV, being previously dubbed as "gay cancer" (Washer, 2014) among homophobic people. The start of the HIV/AIDS epidemic in the United States is attributed largely to the lack of knowledge and research of this new and emerging disease, and the media coverage of what was then termed the "gay pneumonia" (Rayner, 2019). Past research has established an unequivocal association between internalized homophobia, and HIV/AIDS risk behaviors or negative attitudes towards HIV/AIDS prevention efforts among homosexual men (Rayner, 2019). Despite much more accurate findings that the disease also affects individuals of other populations such as intravenous drug users, the association between HIV/AIDS and homosexuality remained dominant in labels such as "gay-related immune deficiency" or GRID. This stigmatization of using the impacts of HIV/AIDS on gay men as an excuse for homophobia, masking it under the idea that it is science can have severe impact for the community. These types of HIV/AIDS-related stigma have been found to be a barrier to regular HIV testing among homesexual men, demonstrating the severe impacts of promoting these false misconceptions (Rayner, 2019). This is not just the impact of homophobia, but also media and news coverage spreading misinformation, and the lack of accurate research of the virus.

This exact case of spreading false information impacting minority groups can be identified in light of COVID-19 as well, establishing a pattern in virus discourse. In the beginning of the pandemic, one would log onto Twitter to see "#ChinaVirus" and "#WuhanVirus' ' trending (Vasquez, 2020). These misinterpretations arise from attaching the origin location and thus ethnicity to the disease. Stereotypes and stigmas would even be expressed to people who have recently travelled to the areas where COVID-19 is spreading. Similar to the case of HIV, the, stigma associated with referring to an illness in a way that deliberately creates unconscious (or conscious) bias, can prevent people from getting care they may desperately need to get better and prevent others from getting sick (Vasquez, 2020). In both cases of HIV and COVID-19, evidence is demonstrating the impact that stigmatizing viruses has severe and dangerous impact on certain demographics mental and physical health. Yet, we still witness this pattern of associating high-risk groups or originated locations and ethinicies to stereotypes and discrimination.

Conclusion

Virus controversy takes on many forms, however in any form mentioned in this chapter the impact proves to be severe. The lack of preparedness and slow policy responses led to outbreaks and continuation of the COVID-19 spread. Additionally, virus conspiracies have prevented people from taking the proper precautions to protect themselves and others. The undeserved impact on marginalized groups that suffer stigmatization of viruses result in alienation and discrimination in times of suffering and chaos. Thus, additional and accurate research, and the promotion of this correct knowledge, is the solution to many of the controversies that arise in the discussions of viruses.

FUTURE RESEARCH AND APPLICATIONS WITH VIRUSES

NOAH VARGHESE

If we learned anything from the COVID-19 pandemic, it is that learning about viruses as a pathogen is of utmost importance. The novel properties and intrinsic diversity of viruses make it a topic of study in the field of healthcare. The novel properties in question include its ability to undergo rapid mutagenesis, replicate rapidly, insert itself into the genome, and take advantage of cellular machinery in novel ways (Vermisoglou et al., 2020). Therefore, it isn't of any surprise that the field of virology was paving way for new technology even before the pandemic. This chapter will explore current research projects investigating viruses for their properties and regarding their potential application in several fields. The chapter aims to offer a summary of the said projects and encourage readers to investigate some of these topics in their own time.

Novel Sensors and Detection Protocols

To discuss the novel technology used to detect the coronavirus, we must first discuss the background and obstacles with current viral detection strategies. A virus's most defining feature is its microscopic size, where it can have a diameter ranging from 20 to 400 nm (Vermisoglou et al., 2020). Current virus detection usually involves extracting the virus from the sample, that being the infected tissue or a fomite, and then culturing it within a cell line (Zhang et al., 2020). The mechanism of detection depends on the nature of the virus, such as what cell type it infects and how it infects it, but amongst all detection procedures, there are some similarities. For example, to detect influenza, the virus is exposed to the Madin Darby canine kidney A549 or rhesus monkey kidney cell lines (Zhang et al., 2020). Then after several days, it is then exposed to erythrocytes to see if the virus adsorbs to the red blood cells. The adsorption of the viral particles to the erythrocytes can be detected through antibody staining and immunofluorescence microscopy, which involves tagging a fluorophore, a chemical that can emit light under certain conditions, to that of an antibody, which is a biological molecule that binds to the virus. Antibodies can be designed in any means, but they are often naturally created by exposing mice to the virus, which its' immune system will naturally create monoclonal antibodies that have the specificity to bind that specific strain of the influenza virus. Note that since influenza rapidly mutates, the antibodies created from influenza one year will probably not bind to the virus from another year. For Ebola, the gold standard wasn't necessarily reliant on culturing but on PCR, but instead on using Reverse Transcription PCR, which in lament terms involves more genealogy study (Saijo et al., 2006). RT-PCR benefits stem from its ability to amplify low viral titres, thereby making the test incredibly sensitive.

One new detection strategy of many viruses, including that of COVID, is the use of graphene working with biosensors (Vermisoglou et al., 2020). Graphene is a light, structured carbon-based nanomaterial that is also highly electrically and thermally conductive. It can also take on many forms such as a nanotube or carbon dots, although these structures are notably more expensive to create. Their conductive properties make them good biosensors, allowing detection to be portable and fast, both of which are valuable features to have in a detection method during the pandemic. The graphene used in the biosensing takes advantage of its electrochemical properties, where the graphene surface is often doped with other metallic compounds, such as with gold or iron oxides in the case of detecting influenza, to improve said properties. The doping and the innate property of graphene make this detection method very sensitive, however, it does rely on very accurate methods of separating the virus from any impurities (Sadighbayan et al., 2020). That being said, even when time is dedicated to the purification of the virus, the graphene detection protocol is still quicker than most other detection methods and is applicable for various other viruses.

Another detection method that is currently being investigated for its future applications is something called Nuclear Acid Amplification Testing, otherwise known as NAAT (Labs, 2020). This has been used before the pandemic for seasonal influenza but is still being researched extensively. NAAT entails the detection of the genetic material of the virus, even if it exists in small proportions, making the tests extremely sensitive. NAAT is unique in that it isn't defined by a specific procedure, as long as the said procedure results in the amplification of the genetic material. Therefore, NAAT has technically been used for a long time through the means of reverse traction PCR, however, there are investigations into other methods (Labs, 2020). In fact, since the pandemic, there has been an increase in the frequency of NAAT, but also the type of NAATs. Only time will tell where this would lead, but perhaps the next damages by the next pandemic will be minimized

New Vaccination Methods
Just like how different detection methods are being investigated, vaccination is also changing. Vaccines have historically

been used for over a century, even if we didn't understand how they worked (Koirala et al., 2020). Due to vaccination, we have been able to control many illnesses such as rabies, polio, and rubella, but we were also able to eradicate smallpox and rinderpest (a disease found in cattle) in its entirety (Koirala et al., 2020). There have been many vaccine strategies, varying in strength and safety that have been used in the past century and all of which have been employed for different illnesses based on their risk-to-benefit ratio. One type of vaccine strategy uses live attenuated vaccines, which involves exposing the immune system to an active, yet safe, variant of the pathogen to induce a strong immune response if the individual ever encounters the real pathogenic virus (Minor, 2015). Another vaccination strategy involved is the use of an inactive variant of the pathogen, which usually has an inert gene or even lacking the genetic material needed for replication (Xia et al., 2020). Other vaccination strategies make use of injecting the protein or antigens, although these are not so consistent in inducing an immune response even though they are far safer than using a live attenuated vaccine (Koirala et al., 2020) (C. Zhang et al., 2019). However, one vaccination strategy has been getting some attention recently due to its use in the COVID-19 pandemic, that being the mRNA vaccination.

mRNA vaccination therapy involves the injections of an RNA transcript that usually codes for a protein antigen found on the virus or pathogen of choice. The beauty of the mRNA vaccine is that because you are injecting the actual transcript, the host cell is responsible for creating the antigens themselves, which can allow for self-amplification of the protein (C. Zhang et al., 2019). Additionally, mRNA vaccines don't need adjuvants given that the nucleic acids can activate the PAMPS, unlike that of other conventional vaccines. Normally to induce the correct immune response one needs to expose the patient to another molecule that guides the immune reaction by binding to PRPs. For viruses, since we need to induce a viral immune response, we would need to normally inject the person with PAMPs of a virus, which can be a nucleic acid. But since the mRNA, which creates the antigen, is itself a nucleic acid, it makes the process of mRNA vaccine delivery very efficient (Schlake et al., 2012). Finally, mRNA vaccines are highly adaptable and can be used in a variety of ways. For example, the vaccine can be injected directly into the patient in vivo, or it can be put exposed to dendritic cells ex vivo which is then later placed within the patients afterwards (Schlake et al., 2012). Ex vivo is the most common method of mRNA vaccine delivery, although it is time-consuming. Even though mRNA vaccines seem perfect, it still warrants research given that the vaccine can induce side effects that are unknown as of now.

Another area of study for vaccination that involves viruses is using a virus to deliver or administer a vaccine. Ironically given that viruses are efficient packets of information designed to spread, it makes them suitable candidates to hold vaccines in the process of delivery. Of course, to minimize danger to the patient, the virus that is used to carry the vaccine is the bacteriophage, a virus only designed to target bacteria, and therefore cannot breach eukaryotic cells (Harada et al., 2018). These bacteriophages often encode antigen proteins in their shell that mimic the pathogen of interest. Additionally, the concept of using bacteriophages to stimulate an immune response can be taken a step further to stimulate an anti-cancer immune response (Hess & Jewell, 2019). Specifically, phages with self-antigens have been shown to suppress tumour development as long ago as the 1940s (Garg, 2019). Other mechanisms have been identified as to how bacteriophages suppress tumours involve it not showing antigens but instead binding to specific sites found on tumour cells. Therefore, in this mechanism phages aren't necessarily inducing an immune response by showing an antigen, like a conventional vaccine, but does through a mechanism usually seen with adjuvants.

Phage Therapy

Phage therapy refers to the use of bacteriophage to target pathogenic bacteria. Superficially, phage therapy is a viable option for killing bacteria given that they are the natural predators for many pathogenic bacteria and cause no harm to humans (Harada et al., 2018). However, the real power behind these bacteriophages isn't necessarily the short-term effects it has for wiping out infections but more so its implications and effects when combined with other anti-bacterial therapies, such as antibiotics (Principi et al., 2019). To better understand the long-term implications of bacteriophage, it is worthwhile summarizing the biggest obstacle with bacterial infection treatment, that being antibiotic resistance (Principi et al., 2019). Regardless of treatment or what type of antibiotic is used, a population of bacteria exposed to an antibiotic will eventually develop resistance to said drug over time due to natural selection (Colomer-Lluch et al., 2011). To further complicate the scenario, bacteria with a gene resistant to antibacterial can even transfer the resistance gene to other bacteria through conjugation, transformation, or transfection, thereby letting the spread of antibiotic genes spread rapidly amongst a population. This has been the reason why antibiotics have slowly been growing less effective and why "superbugs" have been developing. But what researchers are noting is when using bacteriophages in conjunction with antibiotics, the target bacteria can't develop resistance to both therapies simultaneously (Principi et al., 2019). This is in part due to how the gene space available to prokaryotes (who are already limited by a small gene size), only gives them enough room to counteract phages or antibiotics, but not both. It should be noted that the use of bacteriophages isn't necessarily a new science since they have been first used as an antibacterial agent as early as 1966(Principi et al., 2019). However, with the growing concern of antibiotic resistance, it is only recently that bacteriophages have experimented in clinical trials.

There are three approaches for phage therapy, that being the fixed cocktail, modifiable cocktails, and a phage bank; only a phage bank is personalized to that of an individual (Nikolich & Filippov, 2020). The fixed cocktail involves a mixture of phages that can account for the diversity within a bacterium species. Since a single bacterium species can withhold hundreds of different strains, one phage won't be specific enough to target all bacterium in that taxa, hence why several phages are needed. A modifiable cocktail is a fixed cocktail in addition to other phages based on resistance developing amongst pathogens. The phage bank doesn't use a cocktail of phages but instead involves meticulous research on the pathogen and determining which phage(s) can best target said pathogen. Some examples of phage therapy successfully used in clinical trials include work done on Acinetobacter baumannii, Pseudomonas aeruginosa, and Enterobacteriaceae, all of which are dangerous pathogens that are at the forefront of developing antibiotic resistance (Nikolich & Filippov, 2020). In 2017 the first ever instance of human phage therapy trials has been conducted for pancreatic infections of A. baumanni. There has been some success, however another case study revealed that like most modern medications, phage therapy must be personalized based on the condition of the patient and other circumstances. In clinical trials involving phage therapy for Enterobacteriaceae, which is a family that includes that of E.coli, there have also been recorded instances of it lowering biofilm formation, which is clinically relevant given that the biofilms are responsible for food poisoning (Lajhar et al., 2018). There are still steps needed to be taken to use this consistently in a clinical setting, hence why this interesting topic of discussion warrants further research.

Viruses and Gene Therapy

Apart from the use of viruses to prevent the spread of infection of communicable diseases, viruses have also been shown to be a vector in preventing non-communicable diseases. Non-communicable diseases refer to illnesses that are not solely caused by a pathogen but through environmental exposure and mainly genetic deposition. Examples of non-communicable illnesses include cancer, which was already discussed previously, and central nervous system illnesses, both of which have been a target for virus-based gene therapy (Garg, 2019) (Hocquemiller et al., 2016). Gene therapy refers to the use of gene editing machinery to remove or modify disease-causing alleles. It is widely popular with the discovery of CRISPR CAS-9, which is ironically a bacterial protein designed against bacteriophages, however, Adeno-associated viruses (AAV) have been another vector of gene editing (Hocquemiller et al., 2016). The reason why this is the case is that a trait of many viruses, specifically that of lysogenic viruses, is that they integrate themselves into the host genome at specific sites. In theory, if we were to design a virus that can insert itself into a target gene, it can perform a gene knock-out, thereby silencing the gene. Additionally, if the host cell is missing a vital gene, a gene insertion can be performed by using the virus as the delivery vehicle. This is what is being researched using AAV, and there are lots of promising prospects regarding this technology.

The reason why AAV is considered the primary virus for performing gene editing is primarily because it has been studied extensively, it is a lysogenic latent virus known to infect humans and other vertebrates, and, most importantly, has not been connected to any severe human illnesses (Harada et al., 2018). A virus with high latency, like that of AAV, means that it can easily integrate into the genome of the host and will probably not excise itself out like in other lysogenic viruses (Harada et al., 2018). This ensures that the editing performed on the host will remain stable and not revert until after a long duration. Since AAV is extensively studied, researchers are already aware of the mechanism of virus insertion and its genome structure to the extent that they can edit it with intent (Wang et al., 2019). For example, latency can be established in the AAV by removing the helper virus, allowing it to remain integrated within the host genome (Wang et al., 2019). The therapy is still being researched extensively but it has been shown to affect tumour cell lines and treatment of some neurological illnesses like Parkinson's. There is clear progress in the field, but there also drawbacks in implementing gene therapy in current clinical procedures (Principi et al., 2019). For example, one barrier is that the capsid of the AAV, even after extensive modification, can be recognized by the host immune system and targeted, rendering the gene therapy futile (Principi et al., 2019). Additionally, the biggest barrier to the implementation of AAV gene therapy in regular practice is the cost due to how there are many procedures to correctly transfect and modify the AAV insertion genes. That being said, even though prices are dropping for AAV therapy, many companies are discouraged from investing in discovering new kinds of AAV gene therapy (Principi et al., 2019). This will hopefully change as the prices drop and when some AAV drug therapies start being recognized for their therapeutic potential once used in clinical settings.

Conclusion

In conclusion, there are many avenues of research for investigating the applications of viruses, especially in a medicinal setting. The ones of utmost importance, especially during the COVID-19 pandemic, include virus detection and vaccine administration, which itself involves virology in different ways. Additionally, the use of viruses for other biological reasons such as phage therapy, which can revolutionize the field of medicine by countering antibiotic resistance makes it versatile. Additionally, therapy is not only limited to that of communicable illnesses but non-communicable illnesses with AAV gene therapy. If there is any takeaway from this chapter, it is that the applications of viruses, especially in the field of medicine, holds merits.

REFERENCES

WHAT IS THE HISTORY OF VIRUSES?

Breman, J. (2021). Smallpox Eradication: African Origin, African Solutions, and Relevance for COVID-19. The American Journal of Tropical Medicine and Hygiene, 104(2), 416–421. https://doi.org/10.4269/ajtmh.20-1557

Campos-Outcalt, D. (2021). ACIP recommendations for COVID-19 vaccines-and more. The Journal of Family Practice,70(2), 86;89;92–86;89;92. https://doi.org/10.12788/jfp.0153

Cherry, J. (2004). The chronology of the 2002–2003 SARS mini pandemic. Paediatric Respiratory Reviews, 5(4), 262–269. https://doi.org/10.1016/j.prrv.2004.07.009

Crosbie, E., Einstein, M., Franceschi, S., & Kitchener, H. (2013). Human papillomavirus and cervical cancer. The Lancet (British Edition), 382(9895), 889–899. https://doi.org/10.1016/S0140-6736(13)60022-7

de Sanjosé, S., Brotons, M., & Pavón, M. (2018). The natural history of human papillomavirus infection. Best Practice & Research. Clinical Obstetrics & Gynaecology, 47, 2–13. https://doi.org/10.1016/j.bpobgyn.2017.08.015

Douam, F., & Ploss, A. (2018). Yellow Fever Virus: Knowledge Gaps Impeding the Fight Against an Old Foe. Trends in Microbiology (Regular Ed.), 26(11), 913–928. https://doi.org/10.1016/j.tim.2018.05.012

Dworetzky, M., Cohen, S., & Mullin, D. (2003). Prometheus in Gloucestershire: Edward Jenner, 1749-1823. Journal of Allergy and Clinical Immunology, 112(4), 810–814. https://doi.org/10.1016/S0091-6749(03)02017-7

Fitzsimons, T. (2018). LGBTQ History Month: The early days of America's AIDS crisis. https://www.nbcnews.com/feature/nbc-out/lgbtq-history-month-early-days-america-s-aids-crisis-n919701

Francis, M., King, M., & Kelvin, A. (2019). Back to the Future for Influenza Preimmunity-Looking Back at Influenza Virus History to Infer the Outcome of Future Infections. Viruses, 11(2), 122–. https://doi.org/10.3390/v11020122

Greenspan, J. (2020). The Rise and Fall of Smallpox. https://www.history.com/news/the-rise-and-fall-of-smallpox

Griffin, D., & Oldstone, M. (2009). Measles History and Basic Biology (1st ed. 2009.). Springer Berlin Heidelberg. https://doi.org/10.1007/978-3-540-70523-9

Haelle, T. (2019). Why It Took So Long to Eliminate Measles. https://www.history.com/news/measles-vaccine-disease

History. (2020). Influenza. https://www.history.com/topics/inventions/flu

History. (2021). History of AIDS. https://www.history.com/topics/1980s/history-of-aids

Labadie, T., Batéjat, C., Leclercq, I., & Manuguerra, J. (2020). Historical Discoveries on Viruses in the Environment and Their Impact on Public Health. Intervirology, 63, 17-32. https://doi.org/https://doi.org/10.1159/000511575

Malvy, D., McElroy, A., de Clerck, H., Günther, S., & van Griensven, J. (2019). Ebola virus disease. The Lancet (British Edition), 393(10174), 936–948. https://doi.org/10.1016/S0140-6736(18)33132-5

Mehta, N., Julian, P., Meek, J., Sosa, L., Bilinski, A., Hariri, S., Markowitz, L., Hadler, J. & Niccolai, L. (2012). Human Papillomavirus Vaccination History Among Women With Precancerous Cervical Lesions: Disparities and Barriers. Obstetrics and Gynecology, 119(3), 575–581. https://doi.org/10.1097/AOG.0b013e3182460d9f

Moore, Z., Seward, J., & Lane, J. (2006). Smallpox. The Lancet (British Edition), 367(9508), 425–435. https://doi.org/10.1016/S0140-6736(06)68143-9

Moss, W., & Griffin, D. (2012). Measles. The Lancet (British Edition), 379(9811), 153–164. https://doi.org/10.1016/S0140-6736(10)62352-5

Oldstone, M. (2020). Viruses, plagues, and history: past, present, and future (Third edition.). Oxford University Press.

Shi, L., Sings, H., Bryan, J., Wang, B., Wang, Y., Mach, H., Kosinski, M., Washabaugh, M., Sitrin, R., & Barr, E. (2007). GARDASIL®:

Prophylactic Human Papillomavirus Vaccine Development – From Bench Top to Bed-side. Clinical Pharmacology and Therapeutics, 81(2), 259–264. https://doi.org/10.1038/sj.clpt.6100055

Smith, E. (2014). HPV and cancer: the whole story, warts and all. https://scienceblog.cancerresearchuk.org/2014/09/16/hpv-the-whole-story-warts-and-all/

Staples, J., & Monath, T. (2008). Yellow Fever: 100 Years of Discovery. JAMA: the Journal of the American Medical Association, 300(8), 960–962. https://doi.org/10.1001/jama.300.8.960

Song, B., Yun, S., Woolley, M., & Lee, Y. (2017). Zika virus: History, epidemiology, transmission, and clinical presentation. Journal of Neuroimmunology, 308, 50–64. https://doi.org/10.1016/j.jneuroim.2017.03.001

Sullivan, K. (2021). A Brief History of COVID, 1 Year In. https://www.everydayhealth.com/coronavirus/a-brief-history-of-covid-one-year-in/

Taubenberger, J., & Kash, J. (2010). Influenza virus evolution, host adaptation, and pandemic formation. Cell Host & Microbe, 7(6), 440–451. https://doi.org/10.1016/j.chom.2010.05.009

Volberding, P. (2017). How to Survive a Plague: The Next Great HIV/AIDS History. JAMA: the Journal of the American Medical Association, 317(13), 1298–1299. https://doi.org/10.1001/jama.2017.1325

Wagner, R. R. & Krug, R. M. (2020). Virus. Encyclopedia Britannica. https://www.britannica.com/science/virus

Weaver, S., Costa, F., Garcia-Blanco, M., Ko, A., Ribeiro, G., Saade, G., Shi, P., & Vasilakis, N. (2016). Zika virus: History, emergence, biology, and prospects for control. Antiviral Research, 130, 69–80. https://doi.org/10.1016/j.antiviral.2016.03.010

Wikan, N., & Smith, D. (2016). Zika virus: history of a newly emerging arbovirus. The Lancet Infectious Diseases, 16(7), e119–e126. https://doi.org/10.1016/S1473-3099(16)30010-X

Woolhouse, M., Scott, F., Hudson, Z., Howey, R., & Chase-Topping, M. (2012). Human viruses: discovery and emergence. Philosophical Transactions. Biological Sciences, 367(1604), 2864–2871. https://doi.org/10.1098/rstb.2011.0354

World Health Organization. (2018). Zika Virus. https://www.who.int/news-room/fact-sheets/detail/zika-virus

World Health Organization. (2019). Yellow Fever. https://www.who.int/news-room/fact-sheets/detail/yellow-fever#:~:text=Symptoms%20of%20yellow%20fever%20include,and%20Central%20and%20South%20America.

World Health Organization. (2020). HIV/AIDS.https://www.who.int/news-room/fact-sheets/detail/hiv-aids

World Health Organization. (2021). Ebola Virus Disease. https://www.who.int/news-room/fact-sheets/detail/ebola-virus-disease

Zhu, Z., Lian, X., Su, X., Wu, W., Marraro, G., & Zeng, Y. (2020). From SARS and MERS to COVID-19: a brief summary and comparison of severe acute respiratory infections caused by three highly pathogenic human coronaviruses. Respiratory Research, 21(1), 224–224. https://doi.org/10.1186/s12931-020-01479-w

HOW WERE VIRUSES DISCOVERED?

Artenstein, A. W. (2012). The discovery of viruses: Advancing science and medicine by challenging dogma. International Journal of Infectious Diseases, 16(7), e470–e473. https://doi.org/10.1016/j.ijid.2012.03.005

Artenstein, N. C., & Artenstein, A. W. (2010). The Discovery of Viruses and the Evolution of Vaccinology. In A. W. Artenstein (Ed.), Vaccines: A Biography (pp. 141–158). Springer. https://doi.org/10.1007/978-1-4419-1108-7_9

Best, M., & Neuhauser, D. (2004). Ignaz Semmelweis and the birth of infection control. BMJ Quality & Safety, 13(3), 233–234. https://doi.org/10.1136/qshc.2004.010918

Blevins, S. M., & Bronze, M. S. (2010). Robert Koch and the 'golden age' of bacteriology. International Journal of Infectious Diseases, 14(9), e744–e751. https://doi.org/10.1016/j.ijid.2009.12.003

Bos, L. (1999). Beijerinck's work on tobacco mosaic virus: Historical context and legacy. Philosophical Transactions of the Royal Society of London. Series B, Biological Sciences, 354(1383), 675–685. https://doi.org/10.1098/rstb.1999.0420

Cabrera-Perez, J., Badovinac, V. P., & Griffith, T. S. (2017). Enteric immunity, the gut microbiome, and sepsis: Rethinking the germ theory of disease. Experimental Biology and Medicine, 242(2), 127–139. https://doi.org/10.1177/1535370216669610

Claverie, J.-M., & Abergel, C. (2016). Giant viruses: The difficult breaking of multiple epistemological barriers. Studies in History and Philosophy of Science Part C: Studies in History and Philosophy of Biological and Biomedical Sciences, 59, 89–99. https://doi.org/10.1016/j.shpsc.2016.02.015

Daniel-Ribeiro, C. T., & Lima, M. M. (2020). A morning with Louis Pasteur: A short history of the "clean hands." Cadernos de Saúde Pública, 36, e00068619. https://doi.org/10.1590/0102-311x00068619

Ellis, E. L., & Delbrück, M. (1939). The Growth Of Bacteriophage. The Journal of General Physiology, 22(3), 365–384. https://doi.org/10.1085/jgp.22.3.365

Gest, H. (2004). The discovery of microorganisms by Robert Hooke and Antoni van Leeuwenhoek, Fellows of The Royal Society. Notes and Records of the Royal Society of London, 58(2), 187–201. https://doi.org/10.1098/rsnr.2004.0055

Goldstein, B. D. (2012). John Snow, the Broad Street pump and the precautionary principle. Environmental Development, 1(1), 3–9. https://doi.org/10.1016/j.envdev.2011.12.002

Howard-Jones, N. (1977). Fracastoro and Henle: A re-appraisal of their contribution to the concept of communicable diseases. Medical History, 21(1), 61–68. https://doi.org/10.1017/s0025727300037170

Kannadan, A. (2018). History of the Miasma Theory of Disease. ESSAI, 16(1). https://dc.cod.edu/essai/vol16/iss1/18

Karamanou, M., Panayiotakopoulos, G., Tsoucalas, G., Kousoulis, A. A., & Androutsos, G. (2012). From miasmas to germs: A historical approach to theories of infectious disease transmission. Le Infezioni in Medicina, 20(1), 58–62.

Kokayeff, N. (2012). Dying to be Discovered: Miasma vs Germ Theory. ESSAI, 10(1). https://dc.cod.edu/essai/vol10/iss1/24

Krieger, N. (1992). Re: "Who Made John Snow A Hero?" American Journal of Epidemiology, 135(4), 450–451. https://doi.org/10.1093/oxfordjournals.aje.a116305

Liu, Y.-C., Kuo, R.-L., & Shih, S.-R. (2020). COVID-19: The first documented coronavirus pandemic in history. Biomedical Journal, 43(4), 328–333. https://doi.org/10.1016/j.bj.2020.04.007

Opal, S. M. (2010). A Brief History of Microbiology and Immunology. In A. W. Artenstein (Ed.), Vaccines: A Biography (pp. 31–56). Springer. https://doi.org/10.1007/978-1-4419-1108-7_3

Parke, E. C. (2014). Flies from meat and wasps from trees: Reevaluating Francesco Redi's spontaneous generation experiments. Studies in History and Philosophy of Science Part C: Studies in History and Philosophy of Biological and Biomedical Sciences, 45, 34–42. https://doi.org/10.1016/j.shpsc.2013.12.005

Pesapane, F., Marcelli, S., Nazzaro, G., Pesapane, F., Marcelli, S., & Nazzaro, G. (2015). Hieronymi Fracastorii: The Italian scientist who described the "French disease." Anais Brasileiros de Dermatologia, 90(5), 684–686. https://doi.org/10.1590/abd1806-4841.20154262

Pittet, D., & Allegranzi, B. (2018). Preventing sepsis in healthcare – 200 years after the birth of Ignaz Semmelweis. Eurosurveillance, 23(18). https://doi.org/10.2807/1560-7917.ES.2018.23.18.18-00222

Riedel, S. (2005). Edward Jenner and the History of Smallpox and Vaccination. Baylor University Medical Center Proceedings, 18(1), 21–25. https://doi.org/10.1080/08998280.2005.11928028

Rivers, T. M. (1932). The nature of viruses. Physiological Reviews, 12(3), 423–452. https://doi.org/10.1152/physrev.1932.12.3.423

Rundle, C. W., Presley, C. L., Militello, M., Barber, C., Powell, D. L., Jacob, S. E., Atwater, A. R., Watsky, K. L., Yu, J., & Dunnick, C. A. (2020). Hand hygiene during COVID-19: Recommendations from the American Contact Dermatitis Society. Journal of the American Academy of Dermatology, 83(6), 1730–1737. https://doi.org/10.1016/j.jaad.2020.07.057

Schreiner, S. (2020). Ignaz Semmelweis: A victim of harassment? Wiener Medizinische Wochenschrift, 170(11), 293–302. https://doi.org/10.1007/s10354-020-00738-1

Summers, W. C. (2014). Inventing Viruses. Annual Review of Virology, 1(1), 25–35. https://doi.org/10.1146/annurev-virology-031413-085432

Taylor, M. W. (2014). The Discovery of Bacteriophage and the d'Herelle Controversy. In M. W. Taylor (Ed.), Viruses and Man: A History of Interactions (pp. 53–61). Springer International Publishing. https://doi.org/10.1007/978-3-319-07758-1_4

Tognotti, E. (2011). The dawn of medical microbiology: Germ hunters and the discovery of the cause of cholera. Journal of Medical Microbiology, 60(4), 555–558. https://doi.org/10.1099/jmm.0.025700-0

Tyagi, U., & Barwal, K. C. (2020). Ignac Semmelweis—Father of Hand Hygiene. The Indian Journal of Surgery, 1–2. https://doi.org/10.1007/s12262-020-02386-6

van Helvoort, T. (1994). History of virus research in the twentieth century: The problem of conceptual continuity. History of Science, 32(2), 185–235. https://doi.org/10.1177/007327539403200204

Walker, C. (1925). Germ-Theories Of Transferable Diseases From The Seventeenth Century To The Time Of Pasteur. Science

Progress In The Twentieth Century (1919-1933), 19(75), 443–451. http://www.jstor.org/stable/43428369

WHAT IS THE IMPACT OF VIRUSES? - MACRO SOCIETAL VIEW

3M. (2021). COVID-19 Pandemic - Technical Bulletin. https://multimedia.3m.com/mws/media/1791123O/covid-19-pandemic.pdf.

Apuzzo, M., & Kirkpatrick, D. (2020, April 1). COVID-19 Changed How the World Does

Science, Together. New York Times. https://www.nytimes.com/2020/04/01/world/europe/coronavirus-science-research-cooperation.html.

Artenstein, A. (2020). In Pursuit of PPE. New England Journal of Medicine, 382(46). DOI: 10.1056/NEJMc2010025.

Cook, J. (2020). Therapeutic Goods (Excluded Goods—Hand Sanitisers) Determination 2020.

Department of Health (Australia). https://www.legislation.gov.au/Details/F2020L00340.

Dzyakanava, V., Burningham, K., & Stahl, J. (2010). Virucidal efficacy of topical antiseptics

versus a novel strain of Influenza H1N1. International Journal of Infectious Disease, 248(14). https://doi.org/10.1016/j.ijid.2010.02.2041

Fry, C., Cai, X., Zhang, Y., & Wagner, C. (2020). Consolidation in a crisis: Patterns of international collaboration in early COVID-19 research. PLOS ONE, 15(7). https://doi.org/10.1371/journal.pone.0236307

Government of Canada. 2017, June, 8). HIV/AIDS in developing countries.

https://www.international.gc.ca/world-monde/issues_development-enjeux_developpement/global_health-sante_mondiale/hiv_aids-vih_sida.aspx?lang=eng

HM Revenue & Customs. (2020). Producing hand sanitiser and gel for coronavirus (COVID-

19). https://www.gov.uk/guidance/producing-hand-sanitiser-and-gel-for-coronavirus-covid-19#if-you-want-to-produce-hand-sanitiser.

Jecker, N., Wightman, A., & Diekema, S. (2020). Prioritizing Frontline Workers during the

COVID-19 Pandemic. The American Journal of Bioethics. 20(7), 128-132. doi: 10.1080/15265161.2020.1764140

Lane, N (2015). The Unseen World: Reflections on Leeuwenhoek (1677) "Concerning Little

Animal." Philos Trans R Soc Lond B Biol Sci, 370. https://doi:10.1098/rstb.2014.0344.

Lee, G., & Warner, M. (2006). The impact of SARS on China's human resources: implications for the labour market and level of unemployment in the service sector in Beijing, Guangzhou and Shanghai. The International Journal of Human Resource Management, 17(5), 860-880. DOI: 10.1080/09585190600640919

Liu, S., Sun, J., & Cai, J. (2013). Epidemiological, clinical and viral characteristics of fatal cases of human avian influenza A (H7N9) virus in Zhejiang Province, China. Infection, 67(6), 595-605.

Ma, R. (2008). Spread of SARS and War-Related Rumors through New Media in China.

Communication Quarterly, 56(4), 376-391.

MacIntyre, C. R., Chughtai, A. A., Seale, H., Richards, G. A., & Davidson, P. M. (2014).

Respiratory protection for healthcare workers treating Ebola virus disease (EVD): are facemasks sufficient to meet occupational health and safety obligations?. International journal of nursing studies, 51(11), 1421–1426. https://doi.org/10.1016/j.ijnurstu.2014.09.002

MacKellar, L. (2007). Pandemic influenza: A review. Population and Development Review, 33(3), 429-451.

Matsuishi, K., Kawazoe, A., Imai, H., Ito, A., Mouri, K., Kitamura, N., Miyake, K., Mino, K.,

Isobe, M., Takamiya, S., Hitokoto, H. and Mita, T. (2012). Psychological impact of the pandemic (H1N1) 2009 on general hospital workers in Kobe. Psychiatry and Clinical Neurosciences, 66, 353-360. https://doi.org/10.1111/j.1440-1819.2012.02336.x

Morgan, T. (1994). The Industrial Mobilization of World War II: America Goes to War. Army

History, (30), 31-35. http://www.jstor.org/stable/26304207

Paul, S., & Chowdhury, P. (2020). Strategies for Managing the Impacts of Disruptions During

COVID-19: An Example of Toilet Paper, Global Journal of Flexible Systems Management, 21, 283-293. https://doi.org/10.1007/s40171-020-00248-4

Qureshi, A. I., Chinikar, S., & Shahhosseini, N. (2017). Zika virus disease : From origin to

outbreak. ProQuest Ebook Central https://ebookcentral.proquest.com

Smith, R. (2006). Responding to global infectious disease outbreaks: lessons from SARS on the

role of risk perception, communication and management. Social science & medicine, 63(12), 3113-3123.

Snow, J (1849). On the Mode of Communication of Cholera (PDF). London: John Curchill.

Solis-Moreira, J. (2020, December 15). How did we develop a COVID-19 vaccine so quickly?

Medical News Today. https://www.medicalnewstoday.com/articles/how-did-we-develop-a-covid-19-vaccine-so-quickly.

Topic: Fast-track cities. (n.d.) UNIADS. https://www.unaids.org/en/cities

Topic: HIV treatment. (n.d.) UNAIDS. https://www.unaids.org/en/topic/treatment

Ulansky, E. (2016, June 20). The economics of Zika. The Hill. http://thehill.com/opinion/op-

ed/284177-the-economics-of-zika

United Nation Educational, Scientific and Culture Organization. (n.d.). ZIKA: Information

campaign from UNESCO - WHO - IFRC. Retrieved May 14, 2021. https://en.unesco.org/zikainformationcampaign.

U.S Department of Health and Human Services. (2020). Temporary Policy for Preparation of

Certain Alcohol-Based Hand Sanitizer Products During the Public Health Emergency (COVID-19) Guidance for Industry. https://www.fda.gov/media/136289/download.

Weisberg, N. (2020, April 23). We want everyone who wants a mask to be able to get one':

Volunteer mask makers seek help. CTV Edmonton. https://edmonton.ctvnews.ca/we-want-everyone-who-wants-a-mask-to-be-able-to-get-one-volunteer-mask-makers-seek-help-1.4909570.

Wieland, A., & Durach, C. (2021). Two perspectives on supply chain resilience. Journal of

Business Logistics, 42. doi:10.1111/jbl.12271

Wishnick, E. (2010). Dilemmas of securitization and health risks management in the People's

Republic of China: The cases of SARS and avian influenza. Health Policy and Planning, 25(6), 454-466.

Wong, G., & Leung, T. (2007). Bird flu: lessons from SARS. Paediatric Respiratory Reviews, 8(2), 171-176.

Zagorsky, J. (2016, February 23). How Do We Know the Zika Virus Will Cost the World $3.5

Billion? An economist examines how we put a price tag on Zika and other health catastrophes. Scientific American. https://www.scientificamerican.com/article/how-do-we-know-the-zika-virus-will-cost-the-world-3-5-billion/.

WHY IS IT IMPORTANT TO STUDY VIRUSES?

Bonzanic, L. 2020, March 28. Why Are Viruses So Dangerous?. https://medium.com/@lana.bozanic/why-are-viruses-so-dangerous-cda54a9447d5

Burt, K. 2020, February 12. What Makes Viruses so Dangerous? https://www.megainteresting.com/health/article/what-makes-viruses-so-dangerous-751581509557

Cordingley, M. G. (2017). Viruses: agents of evolutionary invention. Harvard University Press.

Crosta, P. 2017, May 30. What to Know About Viruses. https://www.medicalnewstoday.com/articles/158179

King, A. 2020, August 17. Characteristics that Give Viruses Pandemic Potential. https://www.the-scientist.com/feature/characteristics-that-give-viruses-pandemic-potential-67822

Panno, J. (2011). Viruses: the origin and evolution of deadly pathogens. Facts on File.

Tennant, P., Fermin, G., & Foster, J. E. (2018). Viruses: molecular biology, host interactions, and applications to biotechnology. Academic Press.

WHAT ARE VIRUSES?

Crosta, P. (2017, May 30). Viruses: What are they and what do they do? Medical News Today. https://www.medicalnewstoday .com/articles/158179#combating-viruses.

Lodish , H., Zipursky, S. L., et al. (2000). Section 6.3, Viruses: Structure, Function, and Uses. In A. Berk (Ed.), Molecular Cell Biology (4th ed.). essay, New York: W. H. Freeman. https:// www.ncbi.nlm.nih.gov/books/NBK21523/.

WHAT COMMON VIRUSES AND ANTIVIRAL TREATMENTS ARE THERE IN OUR WORLD TODAY?

Arbeitskreis Blut. (2009). Influenza Virus. Transfusion Medicine and Hemotherapy, 36(1), 32–39. https://doi.org/ 10.1159/000197314

Arbeitskreis Blut. (2016). Human Immunodeficiency Virus (HIV). Transfusion Medicine and Hemotherapy, 43(3), 203–222. https:/ /doi.org/10.1159/000445852

Boncristiani, H. F., Criado, M. F., & Arruda, E. (2009). Respiratory Viruses. Encyclopedia of Microbiology, 500–518. https:// doi.org/10.1016/b978-012373944-5.00314-x

Bouvier, N. M., & Palese, P. (2008). The biology of influenza viruses. Vaccine, 26. https://doi.org/10.1016/j.vac-cine.2008.07.039

CDC. (2019, November 18). Types of Influenza Viruses. Centers for Disease Control and Prevention. https://www.cdc.gov/flu/ about/viruses/types.htm.

Crawford, S. E., Ramani, S., Tate, J. E., Parashar, U. D., Svensson, L., Hagbom, M., … Estes,

M. K. (2017). Rotavirus infection. Nature Reviews Disease Primers, 3(1). https://doi.org/10.1038/nrdp.2017.83

Kemnic, T. R., & Gulick, P. G. (2020, June 23). HIV Antiretroviral Therapy. StatPearls.

https://www.ncbi.nlm.nih.gov/books/NBK513308/

Lakna. (2018, February 15). Difference Between Positive and Negative Sense RNA Virus:

Definition, Protein Synthesis, Replication and Differences. Pediaa.Com. https://pediaa.com/difference-between-positive-and-negative-sense-rna-virus/.

Nguyen, H. H. (2020, December 6). What is the global incidence of influenza? Latest Medical

News, Clinical Trials, Guidelines - Today on Medscape. https://www.medscape.com/answers/219557-3459/what-is-the-global-incidence-of-influenza.

Peteranderl, C., Schmoldt, C., & Herold, S. (2016). Human Influenza Virus Infections. Seminars in Respiratory and Critical Care Medicine, 37(04), 487–500. https://doi.org/10.1055/s-0036-1584801

Ramesh, G., MacLean, A. G., & Philipp, M. T. (2013). Cytokines and Chemokines at the

Crossroads of Neuroinflammation, Neurodegeneration, and Neuropathic Pain. Mediators of Inflammation, 2013, 1–20. https://doi.org/10.1155/2013/480739

Razonable, R. R. (2011). Antiviral Drugs for Viruses Other Than Human Immunodeficiency

Virus. Mayo Clinic Proceedings, 86(10), 1009–1026. https://doi.org/10.4065/mcp.2011.0309

Waymack, J. R., & Sundareshan, V. (2020, September 8). Acquired Immune Deficiency

Syndrome. StatPearls. https://www.ncbi.nlm.nih.gov/books/NBK537293/.

WHAT SCIENCE IS INVOLVED IN STUDYING VIRUSES?

Abramowitz, M., & Davidson, M. W. (n.d.). Numerical Aperture and Resolution.

Retrieved May 12, 2021, from Olympus Life Science: https://www.olympus-lifescience.com/en/microscope-resource/primer/anatomy/numaperture/#:~:text=Airy%20Disk%20-Size%20and%20Resolution&text=The%20smaller%20the%20Airy%20disks,do%20objectives%20of%20lower%20correction.

CDC.gov. (2021, April 2). Genomic Surveillance for SARS-CoV-2 Variants. Retrieved May 13, 2021, from Center for Disease Control and Prevention: https://www.cdc.gov/coronavirus/2019-ncov/cases-updates/variant-surveillance.html

Davidson, M. W. (n.d.). Numerical Aperture and Image Resolution page navigation.

Retrieved May 12, 2021, from Microscopyu: https://www.microscopyu.com/tutorials/imageformation-airyna

Lumen Candela. (n.d.). Positive-Strand RNA Viruses in Animals. Retrieved May 12, 2021, from Lumen: boundless microbiology:

https://courses.lumenlearning.com/boundless-microbiology/chapter/positive-strand-rna-viruses-in-animals/

NCBI. (n.d.). Severe acute respiratory syndrome coronavirus 2 isolate Wuhan-Hu-1, complete genome. Retrieved May 13, 2021, from NCBI: https://www.ncbi.nlm.nih.gov/nuccore/1798174254

Oschner Health. (2020, April 16). What does COVID-19 Testing Look Like? - Inside Our COVID-19 Testing Lab. Retrieved May 13, 2021, from https://www.youtube.com/watch?v=Qfrpfod7rgQ

The Conversation. (2021, March 31). Genomic surveillance: What it is and why we need more of it to track coronavirus variants and help end the COVID-19 pandemic. Retrieved May 13, 2021, from The Conversation: https://theconversation.com/genomic-surveillance-what-it-is-and-why-we-need-more-of-it-to-track-coronavirus-variants-and-help-end-the-covid-19-pandemic-157540

WHAT CONTROVERSY IS THERE SURROUNDING VIRUSES?

Altman, D. (2020). Understanding the US failure on coronavirus—an essay by Drew Altman, The BMJ. doi:https://doi.org/10.1136/bmj.m3417

Babu, G. R., Khetrapal, S., John, D. A., Deepa, R., & Narayan, K. V. (2021). Pandemic preparedness and response To COVID-19 in South Asian countries. International Journal of Infectious Diseases, 104, 169-174. doi:10.1016/j.ijid.2020.12.048

Inglesby, T., Toner, E. (2018). Our lack of pandemic preparedness could prove deadly, Washingtonpost.com.https://link.gale

.com/apps/doc/A554879605/AONE?u=ocul_mcmaster&si-
d=AONE&xid=ff6269a5

Prichard, E. C., & Christman, S. D. (2020). Authoritarianism, conspiracy BELIEFS, gender and COVID-19: Links between individual differences and concern About COVID-19, mask wearing
behaviors, and the tendency to Blame China for the virus. Frontiers in Psychology, 11. doi:10.3389/fpsyg.2020.597671

Tan, R. K. (2018). Internalized homophobia, HIV knowledge, and
HIV/AIDS personal Responsibility Beliefs: Correlates of HIV/
AIDS discrimination among MSM in the context of institutional-
ized stigma. Journal of Homosexuality,66(8), 1082-1103.
doi:10.1080/00918369.2018.1491249

Vazquez, M. (2020, March 12). Calling COVID-19 THE "Wuhan
virus" or "china virus" is inaccurate and xenophobic. Retrieved
May 18, 2021, from https://medicine.yale.edu/news-article/
calling-covid-19-the-wuhan-virus-or-china-virus-is-inaccurate-
and-xenophobic/

Washer, P. (2014). Emerging infectious diseases and society. New
York, NY: Palgrave Macmillan.

Wood, M. J. (2018). Propagating and debunking conspiracy
theories on twitter during the 2015–2016 zika virus outbreak.
Cyberpsychology, Behavior, and Social Networking, 21(8), 485-
490. doi:10.1089/cyber.2017.0669

FUTURE RESEARCH AND APPLICATIONS WITH VIRUSES

Colomer-Lluch, M., Jofre, J., & Muniesa, M. (2011). Antibiotic
Resistance Genes in the Bacteriophage DNA Fraction of Environ-

mental Samples. PLoS ONE, 6(3), e17549. https://doi.org/
10.1371/journal.pone.0017549

Garg, P. (2019). Filamentous bacteriophage: A prospective platform for targeting drugs in phage-mediated cancer therapy.
Journal of Cancer Research and Therapeutics, 15(8), 1. https://
doi.org/10.4103/jcrt.jcrt_218_18

Harada, L. K., Silva, E. C., Campos, W. F., Del Fiol, F. S., Vila, M.,
Dąbrowska, K., Krylov, V. N., & Balcão, V. M. (2018). Biotechno-
logical applications of bacteriophages: State of the art. Microbi-
ological Research, 212–213, 38–58. https://doi.org/10.1016/j.mi-
cres.2018.04.007

Hess, K. L., & Jewell, C. M. (2019). Phage display as a tool for
vaccine and immunotherapy development. Bioengineering &
Translational Medicine, 5(1). https://doi.org/10.1002/
btm2.10142

Hocquemiller, M., Giersch, L., Audrain, M., Parker, S., & Cartier,
N. (2016). Adeno-Associated Virus-Based Gene Therapy for CNS
Diseases. Human Gene Therapy, 27(7), 478–496. https://doi.org/
10.1089/hum.2016.087

Koirala, A., Joo, Y. J., Khatami, A., Chiu, C., & Britton, P. N. (2020).
Vaccines for COVID-19: The current state of play. Paediatric
Respiratory Reviews, 35, 43–49. https://doi.org/10.1016/
j.prrv.2020.06.010

Labs. (2020, February 11). Centers for Disease Control and Prevention. https://www.cdc.gov/coronavirus/2019-ncov/lab/
naats.html

Lajhar, S. A., Brownlie, J., & Barlow, R. (2018). Characterization of
biofilm-forming capacity and resistance to sanitizers of a range
of E. coli O26 pathotypes from clinical cases and cattle in Aus-

tralia. BMC Microbiology, 18(1). https://doi.org/10.1186/s12866-018-1182-z

Minor, P. D. (2015). Live attenuated vaccines: Historical successes and current challenges. Virology, 479–480, 379–392. https://doi.org/10.1016/j.virol.2015.03.032

Nikolich, M. P., & Filippov, A. A. (2020). Bacteriophage Therapy: Developments and Directions. Antibiotics, 9(3), 135. https://doi.org/10.3390/antibiotics9030135

Principi, N., Silvestri, E., & Esposito, S. (2019a). Advantages and Limitations of Bacteriophages for the Treatment of Bacterial Infections. Frontiers in Pharmacology, 10. https://doi.org/10.3389/fphar.2019.00513

Sadighbayan, D., Hasanzadeh, M., & Ghafar-Zadeh, E. (2020). Biosensing based on field-effect transistors (FET): Recent progress and challenges. TrAC Trends in Analytical Chemistry, 133, 116067. https://doi.org/10.1016/j.trac.2020.116067

Saijo, M., Niikura, M., Ikegami, T., Kurane, I., Kurata, T., & Morikawa, S. (2006). Laboratory Diagnostic Systems for Ebola and Marburg Hemorrhagic Fevers Developed with Recombinant Proteins. Clinical and Vaccine Immunology, 13(4), 444–451. https://doi.org/10.1128/cvi.13.4.444-451.2006

Schlake, T., Thess, A., Fotin-Mleczek, M., & Kallen, K. J. (2012). Developing mRNA-vaccine technologies. RNA Biology, 9(11), 1319–1330. https://doi.org/10.4161/rna.22269

Vermisoglou, E., Panáček, D., Jayaramulu, K., Pykal, M., Frébort, I., Kolář, M., Hajdúch, M., Zbořil, R., & Otyepka, M. (2020). Human virus detection with graphene-based materials. Biosensors and Bioelectronics, 166, 112436. https://doi.org/10.1016/j.bios.2020.112436

Wang, D., Tai, P. W. L., & Gao, G. (2019). Adeno-associated virus vector as a platform for gene therapy delivery. Nature Reviews Drug Discovery, 18(5), 358–378. https://doi.org/10.1038/s41573-019-0012-9

Xia, S., Duan, K., Zhang, Y., Zhao, D., Zhang, H., Xie, Z., Li, X., Peng, C., Zhang, Y., Zhang, W., Yang, Y., Chen, W., Gao, X., You, W., Wang, X., Wang, Z., Shi, Z., Wang, Y., Yang, X., . . . Yang, X. (2020). Effect of an Inactivated Vaccine Against SARS-CoV-2 on Safety and Immunogenicity Outcomes. JAMA, 324(10), 951. https://doi.org/10.1001/jama.2020.15543

Zhang, C., Maruggi, G., Shan, H., & Li, J. (2019). Advances in mRNA Vaccines for Infectious Diseases. Frontiers in Immunology, 10. https://doi.org/10.3389/fimmu.2019.00594

Zhang, N., Wang, L., Deng, X., Liang, R., Su, M., He, C., Hu, L., Su, Y., Ren, J., Yu, F., Du, L., & Jiang, S. (2020). Recent advances in the detection of respiratory virus infection in humans. Journal of Medical Virology, 92(4), 408–417. https://doi.org/10.1002/jmv.25674

www.ingramcontent.com/pod-product-compliance
Lightning Source LLC
Chambersburg PA
CBHW021823190326
41518CB00007B/720